Synchronization: Theory and Application

NATO Science Series

A Series presenting the results of scientific meetings supported under the NATO Science Programme.

The Series is published by IOS Press, Amsterdam, and Kluwer Academic Publishers in conjunction with the NATO Scientific Affairs Division

Sub-Series

I.   Life and Behavioural Sciences   IOS Press
II. Mathematics, Physics and Chemistry   Kluwer Academic Publishers
III. Computer and Systems Science   IOS Press
IV. Earth and Environmental Sciences   Kluwer Academic Publishers
V.   Science and Technology Policy   IOS Press

The NATO Science Series continues the series of books published formerly as the NATO ASI Series.

The NATO Science Programme offers support for collaboration in civil science between scientists of countries of the Euro-Atlantic Partnership Council. The types of scientific meeting generally supported are "Advanced Study Institutes" and "Advanced Research Workshops", although other types of meeting are supported from time to time. The NATO Science Series collects together the results of these meetings. The meetings are co-organized bij scientists from NATO countries and scientists from NATO's Partner countries – countries of the CIS and Central and Eastern Europe.

Advanced Study Institutes are high-level tutorial courses offering in-depth study of latest advances in a field.
Advanced Research Workshops are expert meetings aimed at critical assessment of a field, and identification of directions for future action.

As a consequence of the restructuring of the NATO Science Programme in 1999, the NATO Science Series has been re-organised and there are currently Five Sub-series as noted above. Please consult the following web sites for information on previous volumes published in the Series, as well as details of earlier Sub-series.

http://www.nato.int/science
http://www.wkap.nl
http://www.iospress.nl
http://www.wtv-books.de/nato-pco.htm

Series II: Mathematics, Physics and Chemistry – Vol. 109

Synchronization: Theory and Application

edited by

Arkady Pikovsky

University of Potsdam,
Potsdam, Germany

and

Yuri Maistrenko

Institute of Mathematics,
National Academy of Sciences of Ukraine,
Kyiv, Ukraine

Kluwer Academic Publishers

Dordrecht / Boston / London

Published in cooperation with NATO Scientific Affairs Division

Proceedings of the NATO Advanced Study Institute on
Synchronization: Theory and Application
Yalta, Crimea, Ukraine
May 19–June 1, 2002

A C.I.P. Catalogue record for this book is available from the Library of Congress.

ISBN 1-4020-1416-3 (HB)
ISBN 1-4020-1417-1 (PB)

Published by Kluwer Academic Publishers,
P.O. Box 17, 3300 AA Dordrecht, The Netherlands.

Sold and distributed in North, Central and South America
by Kluwer Academic Publishers,
101 Philip Drive, Norwell, MA 02061, U.S.A.

In all other countries, sold and distributed
by Kluwer Academic Publishers,
P.O. Box 322, 3300 AH Dordrecht, The Netherlands.

Printed on acid-free paper

Printed in the Netherlands.

Contents

Preface

The NATO Advanced Study Institute on "Synchronization: Theory and Application" was held at the Hotel "Mellas", Yalta Region, Crimea, from 20-31 May 2002.

The topics discussed at the Institute were all concerned with effects of synchrony in the complex dynamics of nonlinear systems. Examples ranged from communication systems through to neuron ensembles. Almost all aspects of the studies of synchronization have been presented: basic mathematical theory, numerical simulations of complex systems, applications of methods of theoretical physics, experimental realizations and applications in engineering and life sciences.

The main feature of the Institute was the three lectures given each day by invited lecturers. In addition, round tables on focused topics have been organized. The lively discussions attested to the enthusiasm and interest that the lecturers succeeded in generating amongst the students. Around a ninety participants from twenty countries, that attended the ASI, have created stimulating atmosphere both inside and outside the lecture room.

We are grateful to our colleagues on the Organizing Committee, Profs. P. Ashwin, M. Hasler, R. Livi and E. Mosekilde for their advice and encouragement and to NATO Scientific Affairs Division for its generous support of the Institute. Additional financial support has been received from the University of Potsdam and the National Academy of Sciences of Ukraine. Last, but not least, we are indebted to our colleagues at the Institute of Mathematics of the National Academy of Sciences of Ukraine as well as the staff of the Hotel "Mellas" without whose dedicated help the organization of the Institute would not have been possible.

A. Pikovsky and Yu. Maistrenko

Contributors

Peter Ashwin — School of Mathematical Sciences, Laver Building, University of Exeter, Exeter EX4 4QE, UK

Alexander Dmitriev — Institute of Radioengineering and Electronics, Russian Academy of Sciences, Mokhovaya St., 11, Moscow, Russia

Martin Hasler — Laboratory of Nonlinear Systems, Swiss Federal Institute of Technology Lausanne, EL-E, EPFL-I&C-LANOS, CH-1015 Lausanne, Switzerland

Kunihiko Kaneko — Department of Pure and Applied Sciences, College of Arts and Sciences, University of Tokyo, Komaba, Meguro-ku, Tokyo 153, Japan

Sergey Kuznetsov — Saratov Division of Institute of Radio-Engineering and Electronics, Russian Academy of Sciences, Zelenaya 38, Saratov, 410019, Russia

Yuri Maistrenko — Institute of Mathematics, National Academy of Sciences of Ukraine, Tereshchenkivska St. 3, 01601 Kyiv, Ukraine

Erik Mosekilde — Department of Physics, The Technical University of Denmark, 2800 Kgs. Lyngby, Denmark

Edward Ott — Institute for Research in Electronics and Applied Physics, Department of Physics, and Department of Electrical and Computer Engineering, University of Maryland, College Park, Maryland, 20742, USA

Arkady Pikovsky — Department of Physics, Potsdam University, Am Neuen Palais 19, PF 601553, D-14415, Potsdam, Germany

Kestutis Pyragas — Semiconductor Physics Institute and Vilnius Pedagogical University, Vilnius, Lithuania

CYCLING ATTRACTORS OF COUPLED CELL SYSTEMS AND DYNAMICS WITH SYMMETRY

PETER ASHWIN*
School of Mathematical Sciences, Laver Building, University of Exeter, Exeter EX4 4QE, UK

ALASTAIR M. RUCKLIDGE †
Department of Applied Mathematics, University of Leeds, Leeds LS2 9JT, UK

ROB STURMAN‡
Department of Applied Mathematics, University of Leeds, Leeds LS2 9JT, UK

Abstract

Dynamical systems with symmetries show a number of atypical behaviours for generic dynamical systems. As coupled cell systems often possess symmetries, these behaviours are important for understanding dynamical effects in such systems. In particular the presence of symmetries gives invariant subspaces that interact with attractors to give new types of instability and intermittent attractor. In this paper we review and extend some recent work (Ashwin, Rucklidge and Sturman 2002) on robust non-ergodic attractors consists of cycles between invariant subspaces, called 'cycling chaos' by Dellnitz et al. (1995).

By considering a simple model of coupled oscillators that show such cycles, we investigate the difference in behaviour between what we call *free-running* and *phase-resetting* (discontinuous) models. The difference is shown most clearly when observing the types of attractors created when an attracting cycle loses stability at a resonance. We describe both scenarios

* P.Ashwin@ex.ac.uk
† A.M.Rucklidge@leeds.ac.uk
‡ rsturman@amsta.leeds.ac.uk

A. Pikovsky and Y. Maistrenko (eds.), Synchronization: Theory and Application, 5–23.

– giving intermittent *stuck-on* chaos for the free-running model, and an infinite family of periodic orbits for the phase-resetting case. These require careful numerical simulation to resolve quantities that routinely get as small as 10^{-1000}.

We characterise the difference between these models by considering the rates at which the cycles approach the invariant subspaces. Finally, we demonstrate similar behaviour in a continuous version of the phase-resetting model that is less amenable to analysis and raise some open questions.

1. Introduction

To understand more complex dynamical systems, it is often helpful to break them down into a number of smaller units or 'cells' that interact with each other. These cells may be imposed naturally by the system one is modelling (for example, neuronal activity), or may just be mathematically helpful (such as linear spatial modes in a spatially extended nonlinear system). Isolating the interactions between the units and the dynamics of the individual units gives a coupled cell description for the dynamics. A basic question for such systems is whether the attracting dynamics of the system is synchronized in any sense.

In cases where the cells are identical the dynamics is constrained by the existence of invariant (synchronized) subspaces for the dynamics. The paper reviews some recent work that exploits symmetries of coupled identical cell systems to help understand their generic behaviour.

The paper proceeds as follows. In section 2 we review some basic concepts from dynamical systems with finite symmetry group and the effects of invariant subspaces. In section 2.1 we discuss the stability and bifurcation of attractors in and near invariant subspaces; a common feature of such attractors is that they may be highly intermittent [2].

In section 2.2 we discuss a class of more complicated intermittent attractors that nevertheless *robustly* appear, that involve a number of invariant subspaces. These attractors have dynamics that show 'cycling chaos' [12] between a number of invariant sets that may be chaotic or periodic. These attractors may be non-ergodic, namely there are obstructions to convergence of averages of observations made on the system.

For the remainder of this article we focus on a particular family of coupled systems introduced in [5] that have robust cycling attractors. Section 3 introduces these cyclically coupled logistic maps, and if we ensure (by introducing a discontinuity in the map) that the approach to consecutive chaotic saddles is via a single trajectory (we call this phase-resetting) we can investigate how their instability causes the appearance of an infinite family of nearby periodic orbits. Finally we present in section 3.5 evidence that phase-resetting can appear even if the map remains smooth.

2. Dynamics with symmetry

Rich dynamics are frequently found in systems that commute with a group of symmetries. These symmetries constrain what can happen in the system while also causing atypical behaviour to become generic. At the simplest level, symmetries cause multistability of attractors; any symmetric image of an attractor must also be an attractor. Moreover, if an initial condition has a certain symmetry, this symmetry must be retained along the trajectory giving rise to invariant subspaces for the dynamics. Symmetries can also cause constrain instabilities by forcing repetition of eigenvalues or Lyapunov exponents corresponding to perturbations that are in symmetrically related directions.

The development of equivariant dynamical systems, or dynamical systems with symmetry, has made great progress in giving a number of tools from group representation and singularity theory to classify such behaviour. This has been particularly successful in classifying local bifurcations of equilibria and periodic solutions (see Golubitsky et al. [14, 15, 16]) but in this article we do not discuss this or its application to coupled cell networks.

2.1. ATTRACTORS IN INVARIANT SUBSPACES AND INTERMITTENCY

For dynamics that is symmetric under linear actions of a finite group, there is a linear invariant subspace associated with each subgroup of symmetries. When an attractor in an invariant subspace loses transverse stability due to a change of parameter we have a so-called *blowout* bifurcation. More precisely, consider a dynamical system on $\mathbb{R}^n$, containing an subspace M of dimension $m < n$ which is invariant under the action of the system. Suppose that for parameters μ below a critical parameter μ_c the subspace M contains a chaotic attractor A of the full system. A blowout bifurcation occurs at μ_c if A ceases to be an attractor for $\mu > \mu_c$. Two different types of blowout bifurcation scenario were characterised in [23] and applications to coupled systems noted in [3]. The first scenario is *subcritical* (also called *hysteretic* or *hard*) and occurs when there are no nearby attractors beyond the bifurcation. This class of blowout is characterised by a riddled basin of attraction. The second is the *supercritical* (also called *nonhysteretic* or *soft* case), in which on-off intermittent attractors branch from the original attractor when the bifurcation parameter increases; for a review of intermittency effects see [2].

Despite the complexity and diversity of the dynamics in such cases, the resulting attractors are generally observed to have ergodic natural measures as far as one can tell from numerical simulations. This seems is true for generic attractors that arise in dynamical systems and means that average quantities (such as Lyapunov exponents) can be computed from attracted trajectories and are natural in the sense that they are the same for almost all initial conditions. One of the most surprising results from the study of symmetric systems is that attractors without ergodicity can be found in fairly simple symmetric systems and moreover these can be robust to (symmetry-respecting) perturbations. A review of these robust heteroclinic cycles can be found in [21] and a classification of more general heteroclinic networks that may arise in symmetric systems can be found in [4].

2.2. CYCLING ATTRACTORS

One of the best-known non-ergodic attractors is the structurally stable heteroclinic cycle between fixed points in $\mathbb{R}^3$ (the "Guckenheimer and Holmes cycle", [18]). This is given by the equation

$$\dot{x} = x(l + ax^2 + by^2 + cz^2)$$
$$\dot{y} = y(l + ay^2 + bz^2 + cx^2) \qquad (1)$$
$$\dot{z} = z(l + az^2 + bx^2 + cy^2).$$

For the system (1) the coordinates planes, the diagonals $x(\pm 1, \pm 1, \pm 1)$ and the axes are all invariant. Moreover, there is an open set of parameters (a, b, c) such that all trajectories off these invariant subspaces approach a robust cycle formed from three equilibrium points on the coordinate axes and trajectories in the coordinate planes connecting these points. Although heteroclinic cycles are possible in systems without symmetry, they are not structurally stable unless there are constraints on the system. The presence of invariant subspaces means that the cycle can be robust – that is, stable with respect to perturbations which preserve invariance of the coordinate planes.

This heteroclinic cycle, illustrated in Figure 1, is robust simply because within the invariant subspaces, the connections are generic connections from saddle to sink. A typical trajectory approaching the cycle will switch between neighbourhoods of the equilibrium points progressively getting closer to the cycle. One can calculate that it will spend a geometrically increasing amount of time close to each equilibrium point and because of this the ergodic averages will be dominated by the present equilibrium and fail to converge. Rather, they will oscillate for ever [25, 19, 13]. On varying a parameter in such a system, several mechanisms whereby robust attracting cycle can be created and destroyed have been identified.

– A *resonance* bifurcation [11] creates a large period periodic orbit at loss of stability of the cycle.

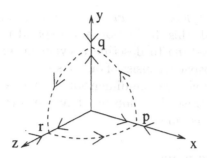

Figure 1. An example of a robust heteroclinic cycle; the Guckenheimer-Holmes cycle in $\mathbb{R}^3$. This is a cycle of heteroclinic connections between the equilibria p, q, r as illustrated, that exists and is attracting for an open set of parameters (a, b, c). The cycle is structurally stable to all perturbations that preserve the coordinate planes; the connection within each coordinate plane is one from a saddle to a sink.

— The fixed points themselves bifurcate in such a way as to destroy the connection, for example at saddle-node or Hopf bifurcations or other transverse bifurcations that may create new longer attracting heteroclinic cycles (Chossat et al. [9, 10]).

Cycles can also exist between sets that are more complicated than simple fixed points but can be created and destroyed in a similar way. For example, if we have cycles between chaotic sets (that is, chaotic saddles – attracting *within* an invariant subspace but repelling in *transverse* directions) we have *cycling chaos*, as discussed in [1, 6, 12], an aspect of a phenomenon referred to as *chaotic itinerancy* by [20]. The stability of such cycles is usually governed by the ratios of Lyapunov exponents at the saddles. Loss of stability can occur at a blowout bifurcation that destroys the set of connections [6] or at a *resonance bifurcation* that corresponds to a resonance of Lyapunov exponents (which occurs when the rates of linear expansion and contraction become equal).

One would like to understand the sort of attractors that are created at a resonance bifurcation. This question was posed and investigated for a specific planar magnetoconvection model with robust cycling chaos in [6]. The cycle in this case was between a chaotic saddle, an equilibrium point and their images under symmetries of the problem. Numerical simulations in [6] suggest that the resonance bifurcation creates a large number of periodic attractors that branch from the cycling chaos. By contrast, for the skew-product example of cycling chaos examined in [1] the resonance was found to give rise not to periodic orbits but to a chaotic attractor with average cycling chaos, or to quasiperiodicity that is intermittent ('stuck on') to the cycling chaos.

An initial attempt to reconcile these differences was made in [5], by introducing the terms *phase-resetting* and *free-running*. For the remainder of this paper we recall this distinction and expand upon the examples in [5]. We use *phase-resetting* to describe a cycle in which the connections between invariant subspaces consist of only a single trajectory (for example, in a flow the connections are one-dimensional, or in a map, they are zero-dimensional). By contrast, *free-running* describes cycles which have a set of many possible connections.

3. Coupled logistic maps

3.1. FREE-RUNNING MODEL

Let $f(x) = rx(1-x)$ denote the logistic map with parameter $r \in [0,4]$ and consider the system introduced in [5]:

$$
\begin{aligned}
x_{n+1} &= f(x_n)e^{-\gamma z_n} \\
y_{n+1} &= f(y_n)e^{-\gamma x_n} \\
z_{n+1} &= f(z_n)e^{-\gamma y_n}.
\end{aligned}
\tag{2}
$$

This map clearly preserves the coordinate planes defined by $xyz = 0$. Three distinct types of evolution are possible for each variable. For example, consider x: if $z \ll 1$ and $x \ll 1$ then x grows approximately linearly – the *growing* phase. For $z \ll 1$ and $x \approx O(1)$, x evolves according to logistic map dynamics – the *active* phase. Finally if $z \approx O(1)$ the dynamics in the x direction is suppressed by the coupling term – the *decaying* phase. For sufficiently large γ we have a robust cycle between invariant sets. In this state, each variable alternates cyclically between the growing, active and decaying phases. We term a change in the phases a *switch*. More precisely, we say a switch occurs when the growing variable exceeds $\ln r/\gamma$. The number of iterations between switches increases approximately geometrically as trajectories approach the invariant subspaces, and this rate of increase depends on the coupling γ. The rate of increase of switching times approaches zero as γ approaches some critical value from above, which forms the limit of the stability of cycling chaos. This geometric increase is examined in more detail in section 3.3.

For $r < 3$ the cycles are between period one points; as r is increased through period doubling we obtain cycles progressively between periodic orbits and then chaotic saddles. Since numerical simulations of this system need to resolve a neighbourhood of the invariant subspaces at high resolution, we use logarithmic coordinates [6, 24]. The time series in Figure 2 is for parameters that produce attracting cycling chaos. This sort of behaviour could be viewed as a sort of antisynchronization; when one variable becomes active, the currently active variables become quiescent. Referring to

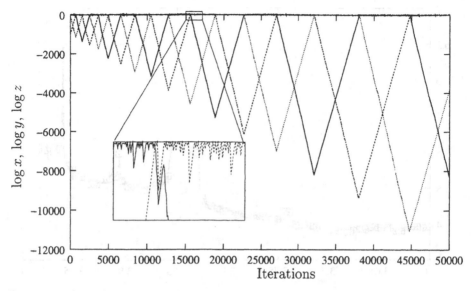

Figure 2.   Attracting cycling chaos, with $r = 4.0$, $\gamma = 6.0$. The model is iterated and x, y, z plotted (in logarithmic coordinates) against time. The chaotic behaviour is $O(1)$ and is visible in the inset. The trajectory cycles through growing, active and then decaying phases for each variable, with the length of phase increasing approximately geometrically. (The same behaviour is found for the phase-resetting version discussed in section 3.3.)

Figure 2, decreasing γ results in a slower rate of increase in the number of iterations between switches, and the line formed by connecting the local minima would become more horizontal.

3.2. COMPUTATION OF RESONANCE BIFURCATION

Suppose that cycling chaos loses stability on decreasing γ through a critical value γ_c. We compute γ_c analytically as follows, as shown in [5]. Suppose that the growing variable is z and a switch has just occurred, so $z \ll 1$, x is $O(1)$ and y is decaying. The evolution of z is governed by $z_{n+1} = rz_n(1 - z_n)e^{-\gamma y_n}$, and this can be approximated by $z \to rz$. Starting at a switch at z_0, suppose that the number of iterations until the next switch is N. Then $z_N \approx r^N z_0$, and since z_N is $O(1)$ at a switch, $N \approx -\ln z_0/\ln r$, where z_0 is the value of z at the start of the growing phase. Whilst z is growing, y is decaying, and for critical γ we require $y_N = z_0$. We approximate y_N in a similar way, with y_0 an $O(1)$ number. Throughout the decay phase $y \ll 1$, but it is forced by the active variable x. Here we approximate by $y \to rye^{-\gamma x}$, and replace x by its long-term average A_∞ $(= \lim_{M \to \infty} \frac{1}{M} \sum_i^M f^i(x_0))$ for each of the N iterations, giving $y_N \approx r^N e^{-\gamma N A_\infty}$. Then substituting our expression for N, we have

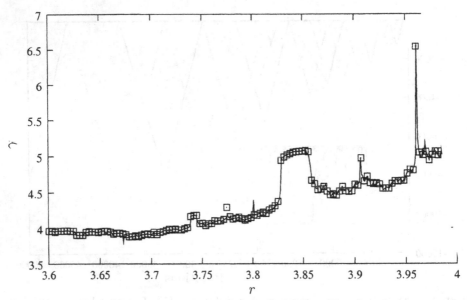

Figure 3. The critical value of γ at which loss of stability of cycling chaos occurs by resonance. The line is estimated using (3) whereas the boxes show points of numerically calculated loss of stability of cycling attractors on varying γ for fixed r. Above the line there is a cycle that is an attractor; below the line the cycling persists but is no longer an attractor.

$\ln y_N \approx -\ln z_0 + (\gamma \ln z_0 A_\infty)/(\ln r)$. The critical value of γ occurs when $\ln y_N = \ln z_0$, giving

$$\gamma_c = 2 \ln r / A_\infty. \tag{3}$$

Equivalently, this can be obtained by considering the ratio of the transversely expanding Lyapunov exponent $\lambda = \ln r$ and the contracting Lyapunov exponent $-\mu = \ln r - \gamma A_\infty$. There will be geometric slowing down with asymptotic rate R, where R is

$$R = \frac{\mu}{\lambda} > 1. \tag{4}$$

There is a resonance when this quantity is equal to unity, also leading to (3).

The average A_∞ is easy to compute numerically, and so we can obtain a curve of critical γ shown in Figure 3, plotted as a line. We superimpose points computed by seeking the parameter at which the number of iterations between switches becomes constant, demonstrating the accuracy of the approximation. When the attracting cycle loses stability in a resonance bifurcation, the resulting stable behaviour is stuck-on chaos – that is, a trajectory which cycles irregularly between invariant subspaces ⸱⸱⸱⸱

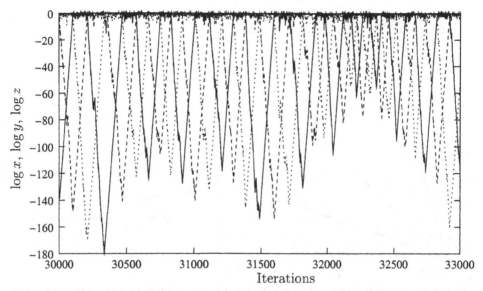

Figure 4.   After losing stability at resonance, in the free-running case, the cycle gives way to a stuck-on attractor. Trajectories are intermittent to each invariant subspace, visiting portions of chaotic behaviour in an irregular fashion but with a finite mean time of cycling. $r = 4.0, \gamma = 5.4$.

portions of chaotic trajectory in an intermittent fashion. A section of such a trajectory is shown in Figure 4.

This agrees with the conjecture in [6] that in a free-running model, cycling chaos loses stability to stuck-on chaos. Although the connections between invariant subspaces here are one-dimensional, because the model deals with discrete time, it is possible for approaches to chaotic saddles to be along many different trajectories. In order to create a phase-resetting version, and thus hopefully obtain periodic orbits, the logistic map was adapted in [5] to force the connections to consist of only a single approach trajectory.

3.3. PHASE-RESETTING MODEL

In order to model the phase-resetting observed in a magnetoconvection system in [6], [5] introduced a phase-resetting enforced by the introduction of a discontinuous 'shelf' in the logistic map. That is, we replace the logistic map f during a growing phase by $\tilde{f}$:

$$\tilde{f}(x_n) = \begin{cases} f(x_n) & x_n < \epsilon \text{ or } x_n > f(\epsilon) \\ f^2(\epsilon) = \eta & x_n \in [\epsilon, f(\epsilon)]. \end{cases}$$

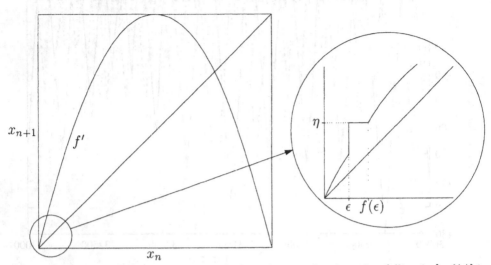

Figure 5. A 'shelf' is introduced to the logistic map. Any iteration falling in $[\epsilon, f(\epsilon)]$ is forced to leave at exactly $\eta = f^2(\epsilon)$. Thus all initial conditions follow the same trajectory into the next saddle.

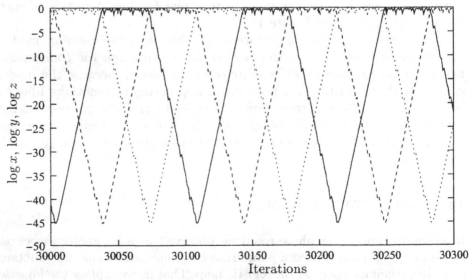

Figure 6. For the phase-resetting case the attracting cycling chaos gives way to many co-existing periodic orbits. The phase-resetting forces each active phase to begin with the same segment of chaotic trajectory. The parameters are $r = 3.75$, $\gamma = 3.9$.

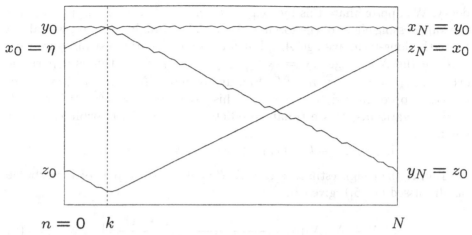

Figure 7. Schematic diagram of a periodic orbit of period $3N$ for the phase-resetting case; one third of a period is shown. This is a periodic orbit as the final and initial phases match up as shown. The iterate k shows where the phases switch.

Each time a growing variable reaches the interval $[\epsilon, f(\epsilon)]$ (we use $\epsilon = 10^{-6}$ in the following), it is set to $\eta = f^2(\epsilon)$, as shown in Figure 5. To ensure that $\epsilon < f(\epsilon)$ we now restrict to $r \in [\frac{1}{1-\epsilon}, 4]$.

Attracting cycling chaos similar to that in Figure 2 can be found for the phase-resetting model and the resonance occurs at the same value of γ_c given by (3) and shown in Figure 3. The only difference from the free-running version is that each portions of chaotic trajectory in the active phases now begin in the same way. The main difference between the two models comes in the behaviour as the cycling loses stability at resonance. In the phase-resetting case, the attracting cycle loses stability not to stuck-on chaos but to many co-existing periodic orbits. One such periodic orbit is shown in Figure 6.

3.4. APPROXIMATION OF THE PERIODIC ORBITS NEAR RESONANCE

The nature of the resetting allows us to predict where periodic orbits are likely to occur without having to compute them using the full three-dimensional map. We do this by considering the evolution of the variables over one third of a periodic orbit of period $3N$ as shown in Figure 7, following the method outlined in [5]. We assume that x has just reset to $x_n = \eta$ at $n = 0$, so that y is in the active phase and z is decaying. For a periodic orbit of period $3N$, we require that $z_N = \eta$ – this will occur when the previous iterate, $z_{N-1} \in [\epsilon, f(\epsilon)]$. We take $y_k = \alpha$, where α is either some $O(1)$ number $\bar{A}$ (for a rough estimate), or more precisely takes the value $f^{N+k}(\eta)$ (since $y_0 = x_N \approx f^N(\eta)$). There follows N iterates of forced decay. We approximate this by $y_{N+k} = r^N y_k e^{-\gamma N\beta}$, where β approximates the suppressing effect of the forcing. Again, for a rough estimate, we take β to be the long-term average A_∞, but for a more accurate estimate we take β to be the N-average $A_N = \frac{1}{N} \sum_{i=0}^{N-1} f^i(f^k(\eta))$. Since this is a periodic orbit, $y_{N+k} = z_k = r^N \alpha e^{-\gamma N\beta}$. Finally we have $(N - k - 1)$ iterates of growth, approximated by $z \to rz$. This gives $z_{N-1} = r^{2N-k-1}\alpha e^{-\gamma N\beta}$. Taking logarithms, this estimate predicts that a periodic orbit can exist when

$$\ln\epsilon < (2N - k - 1)\ln r + \ln\alpha - \gamma N\beta < \ln\epsilon + \ln r.$$

Taking the rough estimates $\alpha = \bar{A}$, $\beta = A_\infty$ gives a pair of hyperbolae (as discussed in [5]), given by

$$N \in [N_1, N_2] = \left[\frac{a}{2\ln r - \gamma A_\infty}, \frac{a + \ln r}{2\ln r - \gamma a_\infty} \right] \tag{5}$$

where $a = \ln\epsilon - \ln\bar{A} + (k+1)\ln r$. We expect periodic orbits to exist for values of N lying between these two hyperbolae. The results in [5] illustrate that this is indeed the case for values of r giving fixed points and periodic solutions within the active invariant subspace, for a suitable choice of fitting parameter $\bar{A}$. Moreover, since the denominators in these expressions equals zero when $\gamma = 2\ln r/A_\infty = \gamma_c$, we can see that the period of the periodic orbits approaches infinity as the coupling parameter approaches the critical value γ_c.

For values of r giving chaotic dynamics within subspaces, the bifurcation diagram of periodic orbits approaching resonance gets more complicated, and the estimate (5) has larger error. We turn to the more sophisticated estimate given by $\alpha = f^{N+k}(\eta)$, $\beta = A_N$ and get

$$z_{N-1} = r^{2N-k-1} f^{N+k}(\eta) e^{-\gamma N A_N}.$$

This is a function only of N (for fixed parameters) and so a curve of z_{N-1} can be easily computed and plotted. Now the method predicts that a periodic orbit can exist for each N for which the curve of z_{N-1} falls within the band given by $[\epsilon, f(\epsilon)]$. The success of this approximation is again illustrated in [5]. Using this method also demonstrates that we expect to find periodic orbits of increasing period as we approach γ_c, and moreover, for the chaotic case, that we may expect periodic orbits to persist even beyond resonance.

Consider the curve of $\ln z_{N-1}$ against N, as illustrated in Figure 8 for values of γ below, equal to, and above γ_c. For $\gamma < \gamma_c$ the fluctuations driven by the NA_N term results in many crossings of the band, but eventually the positive linear average behaviour leads the curve away from the band and the fluctuations are no longer large enough to create more crossings. For $\gamma = \gamma_c$ there is no average increase or decrease, but the mean square fluctuations

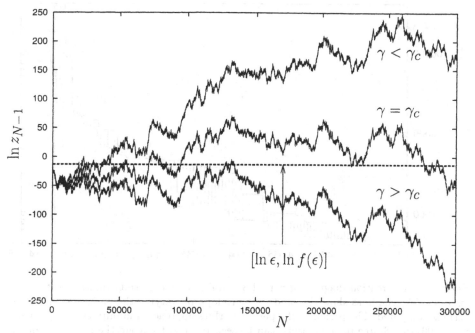

Figure 8. Curves of $\ln z_{N-1}$ given for three different values of γ: a) $\gamma = 4.060 < \gamma_c$, b) $\gamma = 4.061167 = \gamma_c$, c) $\gamma = 4.062 > \gamma_c$.

increase linearly with time in a manner familiar from the central limit theorem. This is to be expected for typical chaotic attractors [7] and means that the curve repeatedly crosses the band, resulting in stable periodic orbits of arbitrarily high period. For $\gamma > \gamma_c$ the negative linear average again leads the curve away from the band, but for values of γ very close to γ_c the possibility of a crossing before the curve is forced too far away remains, and with it the chance of finding periodic orbits persisting beyond resonance.

Figure 9 shows a bifurcation diagram of stable periodic orbits approaching resonance. The lines give the envelope of predicted periodic orbits – the initial and final times the approximation z_{N-1} falls into the band. In between, the dots representing numerically located periodic orbits lie in a complicated structure, but still the approximation works well. In particular just beyond γ_c we see the predicted envelope protrudes past the resonance, and indeed one can locate stable periodic orbits for these parameter values, although most initial conditions appear to lead to cycling chaos instead. This is reminiscent of the phenomenon in Shilnikov-type chaos (for example, see [17]) in which stable horseshoes are observed to exist for parameters on both sides of a homoclinic orbit to a spiral saddle, and also to the similar

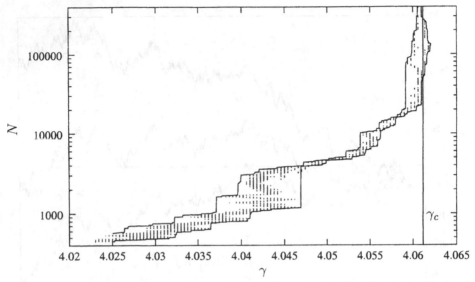

Figure 9. Bifurcation diagram of period $3N$ stable periodic orbits (marked with dots) for $r = 3.75$. The lines show the predicted envelope on varying the parameter γ for the periods computed. The period of the periodic orbits approaches infinity as γ approaches $\gamma_c \approx 4.061165$. Stable periodic orbits can be seen to persist beyond the resonance point.

phenomena observed near cycles to heteroclinic cycles [8].

If we examine the geometrical rate R of increase (4) of switching time as approximated by $R_n = T_{n+1}/T_n$, we observe a difference in behaviour between the free-running and phase-resetting versions of the system. This is illustrated in Figure 10 where we record the ratio of the number of iterations between successive switches. Before resonance, the phase-resetting model exhibits periodic orbits and so as expected (after transients have died down) this ratio tends to unity as the number of iterations between switches becomes constant. (There are also cases where this can tend to a periodically varying function with unit mean in case the periodic orbit modulo the symmetry does not repeat after N iterates but rather after a multiple of N iterates.) The oscillations in the ratio as the periodic orbit is approached were also seen in the flow example of [6]. Conversely, for the free-running case we see the ratio fluctuates as a trajectory follows the irregular cycling of stuck-on chaos.

Beyond resonance, the ratios for both models tend to the same limit R as expected; the convergence appears to be less uniform for the free-running model because of different approaches to the invariant subspaces after each switch; however this is misleading as for example if we choose a value of η for the phase-resetting shelf that has a non-generic orbit this could lead to atypical behaviour for arbitrarily long periods on the cycle.

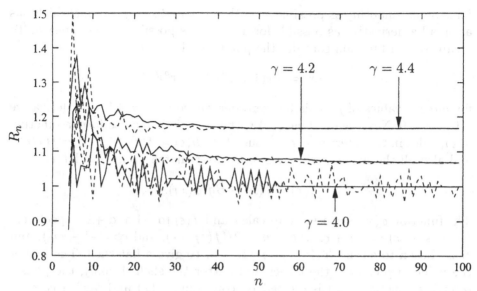

Figure 10. The ratio $R_n = T_{n+1}/T_n$ of number of iterations T_n between switches, n, for the phase-resetting model (solid lines), and the free-running model (dashed lines). The lower pair show a parameter before resonance ($\gamma = 4.0$), giving unity for the periodic orbits in the phase-resetting case, and irregularity for the stuck-on chaos in the free-running model. The other pairs show the free-running model converging more slowly than the phase-resetting model to a geometric increase ($\gamma = 4.2$ and 4.4). The parameter $r = 3.75$ for all pairs.

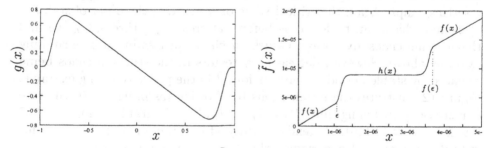

Figure 11. The functions $g(x)$ and $\bar{f}(x)$ for $p = 9$. This is a smoothed version of the 'shelf' resetting in Figure 5.

3.5. SMOOTH MAP WITH PHASE-RESETTING

The map introduced in [5] and discussed in the previous section could be criticized as being degenerate in the phase-resetting case; it has a flat discontinuous 'shelf' introduced to force the growing phase onto a specific trajectory. We show here that it is a straightforward matter to create a similar map that is arbitrarily smooth. For the growing phase we replace the logistic function $f(x)$ within the interval $[\epsilon, f(\epsilon)]$ by a function which joins on as smoothly as possible at both $x = \epsilon$ and $x = f(\epsilon)$, but which has as small a derivative as possible for as much as possible of the interval. To create such a function consider the polynomial

$$g(x) = -x(1 + x^p)^p(1 - x^p)^p$$

for integer values of p. This has zero derivative at $x = \pm 1$, and derivative -1 at $x = 0$. Next we combine $g(x)$ with the logistic map to form a function $h(x)$, valid in the interval $[\epsilon, f(\epsilon)]$, such that $h(\epsilon) = f(\epsilon)$, $h(f(\epsilon)) = f(f(\epsilon))$, and also all the first p derivatives of h and f match up at ϵ and $f(\epsilon)$:

$$h(x) = f(x) + Bg(\phi(x)).$$

The function $\phi(x)$ is chosen to rescale ϵ and $f(\epsilon)$ to -1 and $+1$ respectively (that is, $\phi(x) = mx + c$, where $m = 2/(f(\epsilon) - \epsilon)$, and $c = -1 - m\epsilon$), and B is chosen to ensure $h'((\epsilon + f(\epsilon))/2) = 0$, to give a flat shelf. The higher the value of p chosen, the longer and flatter the shelf. Finally, the phase-resetting function $\tilde{f}(x)$ is created by combining $f(x)$ and $h(x)$ according to

$$\tilde{f}(x_n) = \begin{cases} f(x_n) & x_n < \epsilon \text{ or } x_n > f(\epsilon) \\ h(x) & x_n \in [\epsilon, f(\epsilon)]. \end{cases}$$

The behaviour in this continuous version of the phase-resetting model combines properties from both the free-running and the original phase-resetting maps. Again the critical value of γ marking the onset of stability of cycling chaos can be found as before. Decreasing γ through γ_c we find that cycling chaos gives way to stuck-on chaos, just as in the free-running example above. However, decreasing γ results in the stuck-on chaos being replaced by stable periodic orbits, as found in the phase-resetting example. Figure 12 illustrates these transitions by showing (as in figure 10) the rate of increase of switching times $R_n = T_{n+1}/T_n$. First in graph (a) for $\gamma = 3.87$ (far from $\gamma_c \approx 4.0116$), R_n tends to unity after quite a long transient. This indicates presence of a periodic orbit that closes after a single circuit of the cycle (in this case the period is $N = 35$). Increasing to $\gamma = 4.0$ leaves the system in a state of stuck-on chaos, shown in graph (b) by the ratio fluctuating about an average of unity. Finally in graph (c) the ratio tends to 1.0437 giving, for the post-resonance case $\gamma = 4.2$, the exponent of the geometric increase of switching times associated with cycling chaos.

4. Conclusions

In this work we have briefly reviewed some effects of symmetries on dynamics of coupled cell networks and synchronization. We have also extended [5]

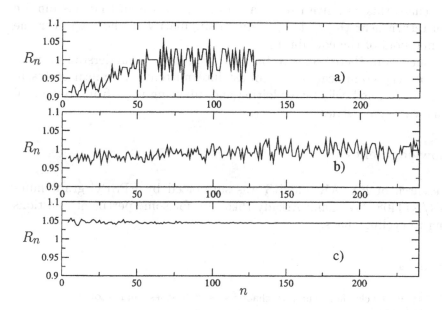

Figure 12. The ratio $R_n = T_{n+1}/T_n$ of number of iterations T_n between switches n for the smooth phase-resetting model. All graphs have $p = 79$ and $r = 3.75$. The top graph a) shows the ratio converging to unity, as a periodic orbit is reached for $\gamma = 3.87$. The middle graph b) shows the irregularity of stuck-on chaos obtained from $\gamma = 4.0$. Finally the bottom graph c) shows the geometric increase of cycling chaos beyond the resonance for $\gamma = 4.15$.

in a number of ways: firstly by considering the ratio of geometric slowing-down for the phase-resetting and free-running cases; and secondly by adapting the phase-resetting case to show that the discontinuity is not essential to give phase-resetting effects.

The model in itself can be interpreted as a ring of mutually-inhibiting cells, and exhibits non-ergodic and intermittent attracting behaviour familiar for robust cycles [21] and 'cycling chaos' [12].

The scenario for loss of stability of a cycle in a flow as investigated in [6] has addition problems in that there is no global section to the attractor and the equilibria are contained within the subspaces that contain the chaotic attractors but nonetheless the maps with singularities appear not to be a bad approximation. In the former model the cycle is formed between alternating saddle equilibria and chaotic saddles, and the phase resetting is caused by the fact that the connection from equilibrium to chaos was along a one-dimensional unstable manifold. This can clearly be robust within an invariant subspace. If we try to make a global section to the flow, near the cycling chaos this will give rise to, at best, a return map that has infinite time of return near the cycle itself and a singularity in the map near the stable manifold of the equilibrium.

In conclusion, there seems to be a lot of promise to understand a wide variety of very complicated but robust intermittent dynamical states in networks of coupled cells by exploiting and adapting tools from dynamical systems with symmetries.

Acknowledgments

The research of PA, AR and RS was supported by EPSRC grant number GR/N14408. We thank Arkady Pikovsky for some pertinent questions relating to cycling chaos.

References

1. P. Ashwin, Cycles homoclinic to chaotic sets; robustness and resonance. *Chaos* **7** 207–220 (1997)
2. P. Ashwin, Chaotic intermittency of patterns in symmetric systems. *Proceedings of IMA workshop on pattern formation in coupled and continuous systems.* Editors: M Golubitsky, D Luss and S H Strogatz, IMA volumes in Mathematics and its applications 115, Springer (1999)
3. P. Ashwin, J. Buescu and I.N. Stewart, Bubbling of attractors and synchronisation of chaotic oscillators, *Physics Letters A* **193** 126-139 (1994)
4. P. Ashwin and M. Field. Heteroclinic networks in coupled cell systems. *Arch. Rational Mech. Anal.* **148** 107–143 (1999)
5. P. Ashwin, A. M. Rucklidge and R. Sturman, Infinities of periodic orbits near robust cycling. In press, *Phys. Rev. E* (2002)
6. P. Ashwin and A. M. Rucklidge, Cycling chaos; its creation, persistence and loss of stability in a model of nonlinear magnetoconvection. Physica D **122** 134-154 (1998)
7. V. Baladi, Decay of correlations, In *Smooth ergodic theory and its applications (Seattle, WA, 1999)* pp 297-325, Amer. Math. Soc., Providence (2001)
8. T. Chawanya, Coexistence of infinitely many attractors in a simple flow. *Physica D* **109** 201–241 (1997)
9. P Chossat, M Krupa, I Melbourne and A Scheel, Transverse bifurcations of homoclinic cycles. Physica D **100** 85-100 (1997)
10. P Chossat, M Krupa, I Melbourne and A Scheel, Magnetic dynamos in rotating convection - A dynamical systems approach. Dynamics of Continuous, Discrete and Impulsive Systems 5 327-340 (1999)
11. S-N Chow, B Deng and B Fiedler, Homoclinic bifurcation at resonant eigenvalues. *J. Dyn. Diff. Eqns.* **2** 177-244 (1990)
12. M. Dellnitz, M. Field, M. Golubitsky, A. Hohmann and J. Ma, Cycling Chaos. *I.E.E.E. Transactions on Circuits and Systems: I. Fundamental Theory and Applications,* **42** 821-823 (1995)

13. A. Gaunersdorfer, Time averages for heteroclinic attractors, *SIAM J. Appl. Math.* **52** 1476–1489 (1992).

14. M. Golubitsky and D. Schaeffer, *Singularities and Groups in Bifurcation Theory, Vol 1.* Springer Applied Math. Sci. vol 51 (1985).

15. M. Golubitsky, I. Stewart and D.G. Schaeffer, *Singularities and Groups in Bifurcation Theory, Vol 2 .* Springer Applied Math. Sci. vol 69 (1988).

16. M. Golubitsky and I. Stewart, *From Equilibrium to Chaos in Phase Space and Physical Space* Birkhauser, Basel, (2002).

17. J. Guckenheimer and P. Holmes, *Nonlinear oscillations, dynamical systems and bifurcations of vector fields*, Applied Mathematical Sciences **42**, Springer Verlag, (1983)

18. J. Guckenheimer and P. Holmes, Structurally stable heteroclinic cycles, *Math. Proc. Camb. Phil. Soc.* **103** 189–192, (1988)

19. J. Hofbauer and K. Sigmund, *The Theory of Evolution and Dynamical Systems*, Cambridge University Press, Cambridge, (1988)

20. K. Kaneko, On the strength of attractors in a high-dimensional system: Milnor attractor network, robust global attraction, and noise-induced selection. *Physica D* **124**, 308–330 (1998)

21. M. Krupa, Robust heteroclinic cycles, *Journal of Nonlinear Science* **7** 129–176 (1997)

22. J. Kurths and S.A. Kuznetsov, *preprint* (2001)

23. E. Ott and J. C. Sommerer, Blowout bifurcations: the occurrence of riddled basins and on-off intermittency. *Phys. Lett. A* **188** 39–47 (1994)

24. A. Pikovsky, O. Popovych and Yu. Maistrenko, Resolving clusters in chaotic ensembles. *Phys. Rev. Lett.*, **87** 044102 (2001)

25. K. Sigmund, Time averages for unpredictable orbits of deterministic systems, *Annals of Operations Research* **37** 217–228 (1992)

MODELLING DIVERSITY BY CHAOS AND CLASSIFICATION BY SYNCHRONIZATION

OSCAR DE FEO* and MARTIN HASLER
Laboratory of Nonlinear Systems
Swiss Federal Institute of Technology Lausanne
EL-E, EPFL-I&C-LANOS, CH-1015 Lausanne, Switzerland

Abstract

A new chaos-based technique for modelling the diversity of approximately periodic signals is introduced and exploited, combined with generalized chaotic synchronization phenomena, for the solution of temporal pattern recognition problems.

1. Introduction

Classification of objects based on examples is one of the central problems of machine learning [26]. In this paper, the objects are approximately periodic signals and the examples are measured time-series. Signals of this kind are abundant in nature, *e.g.* physiological signals like ECG's or EEG's, part of speech signals, seismic activities, sea tides, etc.

The main difficulty of classification is to express the diversity of data that has essentially the same origin without creating confusion with data that has a different origin. This is an old subject and consistent literature is available on it (*e.g.* [18, 19, 1, 4, 5]). At the same time, the problem is of too general a nature to allow for a single satisfactory all-embracing classification method.

Normally, the diversity of time-series belonging to the same class is modelled by a stochastic process, such as filtered white noise, a Hidden Markov Model, or a stochastic differential equation [21, 1, 25]. In the more

---

* Oscar.DeFeo@epfl.ch

A. Pikovsky and Y. Maistrenko (eds.), Synchronization: Theory and Application, 25–40.
© 2003 *Kluwer Academic Publishers. Printed in the Netherlands.*

specific case of approximately periodic signals encountered in nature, it is reasonable to assume that there is an underlying dynamical system that generated them, rather than a stochastic process. Under this assumption, the diversity of the data is expressed by the variability of the parameters of the dynamical system. The parameter variability itself is then, once again, modelled by a stochastic process. Thus, the diversity within a class is generated by some form of exogenous noise.

In this paper a different approach is followed. A single chaotic dynamical system is used to model the time series and its diversity. Indeed, a chaotic system produces a whole family of trajectories that are all different but nonetheless very similar [20, 16]. It is believed that chaotic dynamics not only are a convenient means to represent diversity but that in many cases the origin of diversity actually stems from chaotic dynamics. However, whether or not this is the case is not important for the classification task considered in this paper [7]. It would be an issue, if the model were used to generate synthetic signals belonging to the class.

In this paper the following approximately periodic physiological signals serve as examples: electrocardiograms (ECG's), parts of speech signals, and electroencephalograms (EEG's). Since there are often strong arguments in favor of the chaotic nature of these signals, they appear to be the best candidates for modelling diversity by chaos. It is repeated, however, that this modelling approach is thought to be quite general and whether or not a chaotic system has produced the signals under consideration is not crucial for being able to perform the classification task.

2. Classification Problem

Often, classification has to choosing among a number of classes. For example, reading handwritten digits amounts to choose among the 10 classes labelled by 0, 1, ..., 9. For simplicity, we limit ourselves to the case of only two classes. Thus, two classes of approximately periodic signals are given and we have to find an algorithm that decides to which class a signal belongs. Such a classification problem can be easy or hard, depending on:

1. how "close" the two classes are;
2. how the two classes are defined.

In our case, the classes are defined indirectly by a representative set of examples, in the form of a database of recorded signals, labelled with the class symbol. From the examples, a model for the classes has to be deduced. This operation is called "supervised learning" [22]. In fact, the learning is supervised, because for each recorded signal the class is known, the "teacher" or "supervisor" tells us what the class is.

To fix the ideas, we give two examples. The first concerns vowels in speech recognition. A database of several (about 50) recorded and labeled [a]'s and [e]'s is given and the task consists of distinguishing, in a given speech segment that is supposed to represent either a spoken [a] or a spoken [e], which one of the two vowels actually has been pronounced. An example of a spoken [a] is represented in Fig. 1(a), and an example of a spoken [e] in Fig. 1(b).

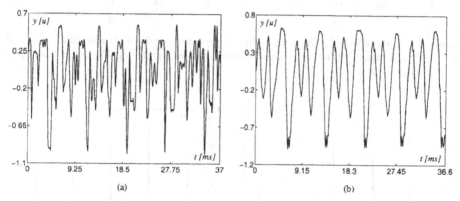

Figure 1.   Spoken vowels: (a) – recorded [a]; (b) – recorded [e].

Since these signals are approximately periodic, they can be decomposed into "pseudo-periods". Since they are not precisely periodic, the pseudo-periods are slightly different and the signals within the various pseudo-periods also differ slightly, even if they are time-aligned. Averaging over the time-aligned signals, one obtains a periodic "generating" cycle for the [a] and for the [e] spoken vowels. The set of time-aligned signal portions are represented in Fig. 2, with a highlighted generating cycle.

The second example concerns electrocardiograms (ECG's). The first class of ECG's signals has been taken from persons that have a certain pathology. An example of such an ECG is given in Fig. 3(a). They have to be distinguished from healthy persons. A corresponding ECG is given in Fig. 3(b).

Again, both healthy and pathological ECG's are approximately periodic signals. The pseudo-periods can be normalized, time-aligned signals can be computed and corresponding waveforms within a pseudo-period can be superposed in order to illustrate the time-variability within a single signal, or the variability among different signals (*cf.* Fig. 4). Also, a periodic generating cycle can be computed.

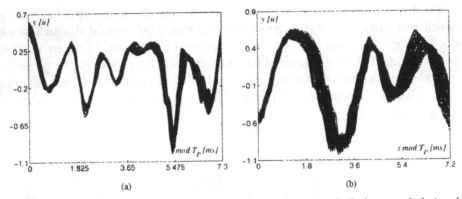

(a)   (b)

Figure 2.   Spoken vowels: (a) – time-aligned pseudo-periods of all the recorded signals [a]; (b) – time-aligned pseudo-periods of all the recorded signals [e]. The generating cycles are represented in bold.

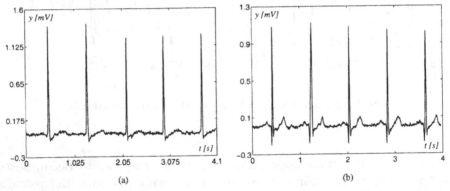

(a)   (b)

Figure 3.   Recorded ECG's: (a) – from a person having a certain pathology; (b) – from a healthy person.

3. Modelling by Nonlinear Dynamic System Identification

Instead of modelling a signal class by a stochastic process, we use a single dynamical system as a model for the whole class [8]. The reasoning goes as follows. The different signals within a class have much similarity, without being identical (*cf.* Figs. 2 and 4). The same is true for the various trajectories within a chaotic attractor of a nonlinear dynamical system as well as for the output signals of such a system, as illustrated in Fig. 5. This leads to the idea to use a chaotic system, or more precisely, the output of a chaotic system, as a model for a signal class. In this way, the diversity of the different signals within a class is represented by the diversity of the output signals that is generated by the diversity of the trajectories in the

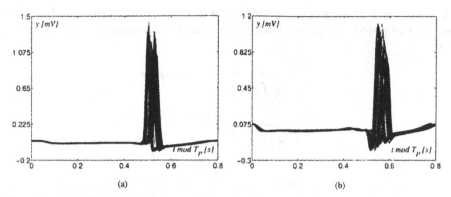

(a)   (b)

Figure 4.   Recorded ECG's: (a) – time-aligned pseudo-periods of all the recorded patho-logical ECG's; (b) – time-aligned pseudo-periods of all the recorded healthy ECG's. The generating cycles are represented in bold.

attractor of the system. In general, an attractor of a dynamical system may be very simple, such as a closed curve or a torus, or rather complicated such as a chaotic attractor. In our application, the diversity of the given signal classes is such that apparently a chaotic attractor is needed for modelling.

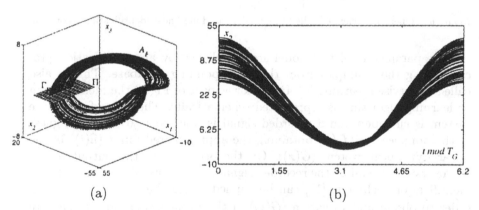

(a)   (b)

Figure 5.   Diversity of the signals generated by a chaotic system: (a) – approximately periodic strange attractor in the state space; (b) – time-aligned pseudo-periods of one of the state variables. The generating cycle is represented in bold.

We choose a Lur'e system [2] as a reference model (upper part of Fig. 6). Its ring structure composed of a nonlinear static dynamic and a linear system has distinct advantage for the modelling process. For computational convenience, we restrict the nonlinearity to be one-dimensional. If it were not for this constraint, the Lur'e systems would actually represent the most general class of finite dimensional nonlinear dynamical systems. To be precise, the 1-dimensional nonlinear function we use is a piecewise linear function composed of 5 pieces, whose angles have been smoothed to second order.

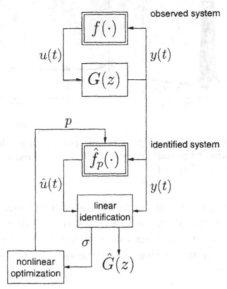

Figure 6. Operating scheme of the Lur'e model based nonlinear identification algorithm.

The parameters of the model are established by an identification process, using the examples from the corresponding database. This is also called supervised learning [4]. In the lower part of Fig. 6 the identification or learning algorithm is represented schematically. The loop of the Lur'e system is cut open and a recorded signal is injected into an initial guess of the nonlinearity ($f_p(\cdot)$) obtaining the approximated input ($\hat{u}(t)$) of the linear dynamical system ($G(z)$). On the other hand, the output of the linear system should be the recorded signal itself. Thus, a parametric linear identification technique [17] can be applied, using the pair ($\hat{u}(t), y(t)$), in order to obtain an estimation ($\hat{G}(z)$) of the linear system and a measure of its quality (σ). This quality measure is then used as a cost function for the optimization of the nonlinearity parameters (p). The procedure is iteratively repeated until the best possible pair ($p, \hat{G}(z)$) is determined. A certain number of constraints has to be applied to the identification process, however, in order to avoid that it converges to the trivial solution, where the loop functions act as the identity operator.

The Lur'e structure of the nonlinear dynamical system has allowed us to use alternated linear and nonlinear system identification. Keeping the nonlinear function fixed, we adjust the linear dynamic part using a standard algorithm. Inversely, keeping the linear part fixed, we adjust the nonlinear function by a genetic algorithm [15, 6]. The result of this identification procedure is shown in Figs. 7 to 10. It can be seen that the synthetic signals produced by the Lur'e systems resemble closely the recorded signals.

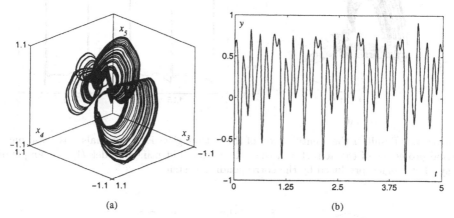

(a)   (b)

Figure 7. Result of the identification of the acoustic signals [a]: (a) – 3-dimensional projection of the attractor of the 5-dimensional identified model; (b) – example of a synthetic signal produced by the corresponding system.

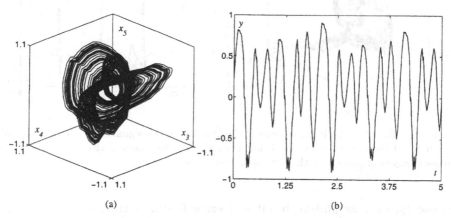

(a)   (b)

Figure 8. Result of the identification of the acoustic signals [e]: (a) – 3-dimensional projection of the attractor of the 5-dimensional identified model; (b) – example of a synthetic signal produced by the corresponding system.

4. Classification by Synchronization

Having obtained a nonlinear dynamical system that autonomously produces signals that resemble closely the recorded signals of the class it

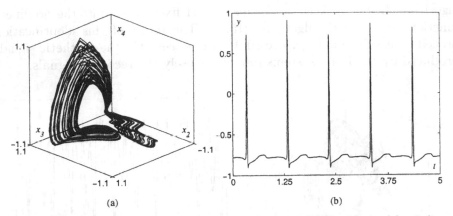

(a)   (b)

Figure 9.   Result of the identification of the pathological ECG's signals: (a) – 3-dimensional projection of the attractor of the 4-dimensional identified model; (b) – example of a synthetic signal produced by the corresponding system.

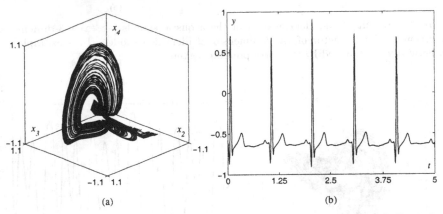

(a)   (b)

Figure 10.   Result of the identification of the healthy ECG's signals: (a) – 3-dimensional projection of the attractor of the 4-dimensional identified model; (b) – example of a synthetic signal produced by the corresponding system.

represents, we now add an input and error feedback to the system (right part of Fig. 11) [14]. Signals that belong to the class that is modelled by the dynamical system should approximately synchronize with the system, whereas signals that do not belong to the class should not synchronize, not even approximately. Thus classifying amounts to checking for approximate synchronization. In this context, the choice of the right feedback coefficients is crucial for achieving a minimal probability of misclassification. As will be discussed in more detail later, too weak feedback will not allow to recognize the signals that belong to the class, whereas too strong feedback will erroneously identify some signals as belonging to the class.

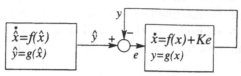

Figure 11.   Master-slave configuration for synchronization controlling the slave by error feedback.

The idea behind this approach to classification is the following [14]. If we connect two identical nonlinear dynamical systems in a master-slave configuration as shown in Fig. 11, then by suitably adjusting the feedback coefficients the slave system will synchronize with the master system. This synchronization is caused by the output signal of the master system alone, because no other information reaches the slave system. Thus, if the output of the master system is recorded, and later replayed at the input of the slave system, the latter will still synchronize with the replayed signal. Now, if the signal at the input of the slave system is only approximately like an output signal from the master system, the slave system will still approximately synchronize with the incoming signal. Hence, if the master system (and thus the slave system without input and feedback) models a signal class then the slave system with input and suitable feedback will approximately synchronize with input signals from the class and not synchronize with other signals.

It turns out [11, 9, 10] that even though the systems that were obtained by our learning/identification process reproduce quite faithfully the signals of their class in the autonomous mode, they are not yet suitable for the classification process. The reason is that their dynamics have a rather rigid underlying approximate periodicity. If the incoming signal is out of phase with the internal dynamics of the system, the feedback will not be able to lock the system onto this signal. Various remedies could be imagined. The remedy we have successfully applied is based on chaos theory. The idea is to modify the system by carefully changing its parameters until it has a homoclinic loop, *i.e.* a very special trajectory that converges for infinite negative and infinite positive time to the same (unstable) equilibrium point [23, 24]. A typical trajectory of the modified system alternates periods of time when it remains close to the equilibrium point and periods when it oscillates in a region of the attractor that is similar to the attractor of the previous, non modified system [12, 13]. The corresponding output signal alternates between almost periodic oscillations similar to the signals the system is supposed to model, and almost constant behavior. Thus, the presence of the homoclinic loop introduces phase-slips into the output signals of the free-running system. When a suitable error feedback is applied

to this modified system, synchronization with an incoming approximately
periodic signal is possible thanks to the phase slips (*cf.* Figs. 12 and 13).
Hence, even though the modified system in the autonomous mode produces
signals that are not so similar to the recorded signals anymore, the system
is much more flexible for synchronizing with a suitable input signal.

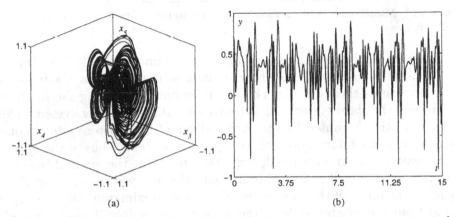

(a)   (b)

Figure 12.   Modified system, to have a homoclinic loop, of the acoustic signals [a]:
(a) – 3-dimensional projection of the attractor of the 5-dimensional modified model (with
homoclinic loop); (b) – example of the corresponding system output signal, comparing
with Fig. 7(b), the phase slips are visible.

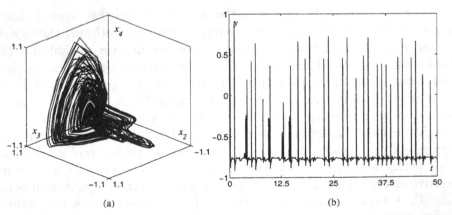

(a)   (b)

Figure 13.   Modified system, to have a homoclinic loop, of the healthy ECG's signals:
(a) – 3-dimensional projection of the attractor of the 4-dimensional modified model (with
homoclinic loop); (b) – example of the corresponding system output signal, comparing
with Fig. 10(b), the phase slips are visible.

As mentioned before, the feedback coefficients in Fig. 11 have to be set
in such a way that approximate synchronization takes place for the signals

of the class the dynamical system models, and no synchronization for the signals of the wrong class. If the feedback is too weak, synchronization hardly ever happens, whereas if the feedback is too strong, the system will synchronize also with wrong signals. The following idea helps to find the right feedback. The crucial trajectories in the attractor of the free-running modified system are the periodic generating cycle and the homoclinic loop. They are represented in Fig. 14 for a system that serves just for illustration purposes.

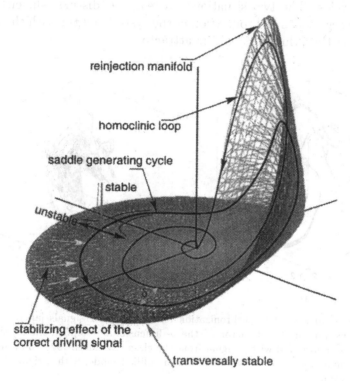

Figure 14. Attractor with a generating cycle and a homoclinic loop, highlighted in bold.

Close to the generating cycle, the dynamics produce output signals similar to those of the class the system models. However, the trajectories of the free-running system explore the whole attractor and therefore they cannot remain close to the generating cycle forever. On the other hand, in the system with input, when a signal of the right class is injected, the feedback control should keep it close to the generating cycle once it enters into its vicinity, whereas a signal of the wrong class should not be captured by the generating cycle. The corresponding feedback coefficients are determined by periodic control theory applied to the system linearized about the generating cycle [3].

In Figures 15(a) and 16(a) the attractors of the modified systems for the acoustic signals [a], and the healthy ECG's signals, with the right input signals are represented, respectively. With respect to the attractors of the autonomous systems shown in Figs. 12 and 13, they are much thinner. Indeed, they are concentrated around their corresponding generating cycles. In Figures 15(b) and 16(b), the attractors of the same systems are shown, when the inputs are the signals from the wrong classes. They clearly fill out much more of the state spaces and they do not stay close to their generating cycles. The two situations are easy to distinguish, either by checking the degree of synchronization of the system output with the input, or by checking the "thickness" of the attractors.

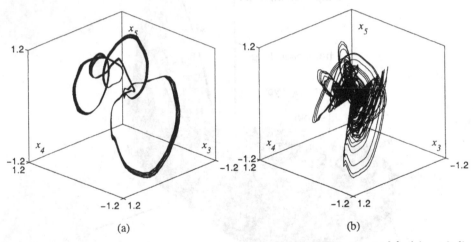

(a)   (b)

Figure 15.   Classification by synchronization for the acoustic signals [a]: (a) – 3-dimensional projection of the attractor of the 5-dimensional modified model that has approximately synchronized with a signal from the class it models; (b) – 3-dimensional projection of the attractor of the 5-dimensional modified model with a signal from the wrong class as input.

The classification results are given in Table I for the examples mentioned above. They are quite reasonable. Classical methods trimmed to the specific application can certainly achieve still better results. However, we have been able to show the feasibility of this entirely different approach.

We also have been able to classify EEG's signals according to whether the person was sleeping or simply drowsy. The corresponding recorded signal are shown in Fig. 17, whilst prototype period of the signals are represented in Fig. 18. It can be seen that the signals within one period are rather complex. Accordingly, the nonlinear dynamical system identification was difficult, but nevertheless we succeeded in obtaining a good models of order 7, whose attractors and output signals are represented in Figs. 19 and

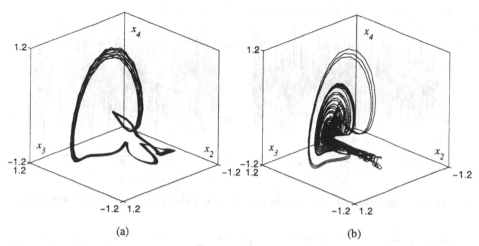

(a)   (b)

Figure 16. Classification by synchronization for the healthy ECG's signals: (a) – 3-dimensional projection of the attractor of the 4-dimensional modified model that has approximately synchronized with a signal from the class it models; (b) – 3-dimensional projection of the attractor of the 4-dimensional modified model with a signal from the wrong class as input.

TABLE I.   Classification results, only the vectors not used for learning are classified: (a) – ECG's signals; (b) – vowels signals.

in \as	P	H
P	88.19%	11.81%
H	14.98%	85.02%

in \as	[a]	[e]
[a]	85.33%	14.67%
[e]	12.62%	87.38%

(a)   (b)

20, respectively. Classification was also successful, but a higher error rate compared with the other examples was observed.

5. Conclusions

We have shown that the diversity of approximately periodic signals found in nature can be modelled by means of chaotic dynamics. Furthermore, we have illustrated how to exploit this kind of modelling technique, together with selective properties of the synchronization of chaotic systems, for pattern recognition purposes.

DE FEO AND HASLER

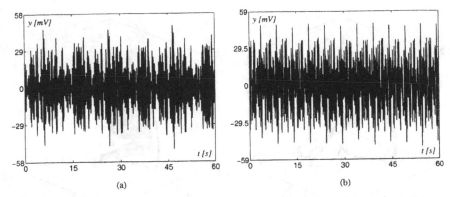

(a)   (b)

Figure 17.   Recorded EEG's: (a) – from a sleeping person; (b) – from a drowsy person.

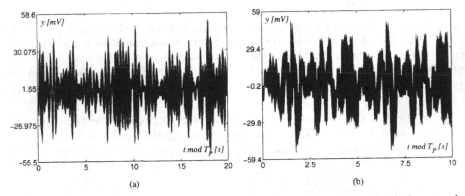

(a)   (b)

Figure 18.   Recorded EEG's: (a) – time-aligned pseudo-periods of all the recorded sleeping EEG's; (b) – time-aligned pseudo-periods of all the recorded drowsy EEG's. The generating cycles are represented in bold.

Acknowledgements

This work was supported by the Swiss National Science Foundation: FN-2000-63789.00; and from the European project APEREST: IST-2001-34893 and OFES-01.0456.

References

1. Alder, M.: 1994, *Principles of Pattern Classification: Statistical, Neural Net and Syntactic Methods of Getting Robots to See and Hear.*   Not published.   Freely available on the world wide web: ftp://ciips.ee.uwa.edu.au/pub/syntactic/book,http://ciips.ee.uwa.edu.au/ mike/PatRec
2. Atherton, D.: 1982, *Nonlinear Control Engineering.* Melburne. Australia: Van Nostrand Reinhold.

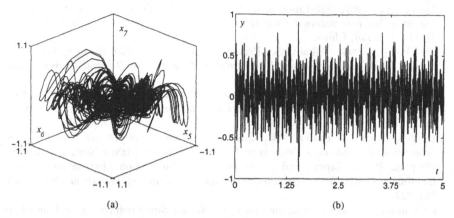

(a)   (b)

Figure 19. Result of the identification of the sleeping EEG's signals: (a) – 3-dimensional projection of the attractor of the 7-dimensional identified model; (b) – example of a synthetic signal produced by the corresponding system.

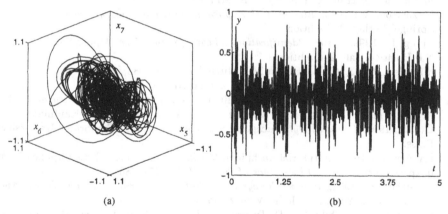

(a)   (b)

Figure 20. Result of the identification of the drowsy EEG's signals: (a) – 3-dimensional projection of the attractor of the 7-dimensional identified model; (b) – example of a synthetic signal produced by the corresponding system.

3.   Bittanti, S. and P. Colaneri: 1999, *Periodic Control*, pp. 59–74. New York, NY: John Wiley & Sons.
4.   Bittanti, S. and G. Picci (eds.): 1996, *Identification, Adaptation, Learning: The Science of Learning Models from Data*. New York, NY: Springer-Verlag.
5.   Cherkassky, V. and F. Mulier: 1998, *Learning from Data: Concepts, Theory and Methods*. New York, NY: John Wiley & Sons.
6.   Dasgupta, D. and D. McGregor: 1994, 'A more biologically motivated genetic algorithm: The model and some results'. *Cybernetics and Systems* **25**, 447–469.
7.   De Feo, O.: 2001, 'Modeling Diversity by Strange Attractors with Application to Temporal Pattern Recognition'. Ph.D. thesis, Swiss Federal Institute of Technology Lausanne, Lausanne, Switzerland.

8.   De Feo, O.: 2002, 'Self-Emergence of Chaos in Identifying Irregular Periodic Be-
     haviors'. In: *International Conference on Nonlinear Theory and its Applications
     NOLTA*. X'ian, China.

9.   De Feo, O.: 2003a, 'Qualitative Resonance of Shil'nikov-like Strange Attractors,
     Part I: Experimental Evidence'. *International Journal of Bifurcation and Chaos*.
     To Appear.

10. De Feo, O.: 2003b, 'Qualitative Resonance of Shil'nikov-like Strange Attractors,
     Part II: Mathematical Analysis'. *International Journal of Bifurcation and Chaos*.
     To Appear.

11. De Feo, O. and M. Hasler: 2001, 'Qualitative resonance of chaotic attractors'. In:
     International Conference Progress in Nonlinear Science. Nizhny Novgorod, Russia.

12. Gaspard, P., R. Kapral, and G. Nicolis: 1984, 'Bifurcation phenomena near ho-
     moclinic systems: A two-parameter analysis'. *Journal of Statistical Physics* **35**,
     697–727.

13. Glendinning, G. and C. Sparrow: 1984, 'Local and global behavior near homoclinic
     orbits'. *Journal of Statistical Physics* **35**, 215–225.

14. Hasler, M.: 1994, *Synchronization Principles and Applications*, pp. 314–327. New
     York, NY: IEEE Press.

15. Koza, J.: 1992, *Genetic Programming: On the Programming of Computers by Means
     of Natural Selection*. Cambridge, MA: MIT Press.

16. Kuznetsov, Y.: 1998, *Elements of Applied Bifurcation Theory*. New York, NY:
     Springer-Verlag, 2nd edition.

17. Ljung, L.: 1999, *System Identification: Theory for the User*. Upper Saddle River,
     NJ: Prentice-Hall, 2nd edition.

18. Michalski, R., J. Carbonell, and T. Mitchell (eds.): 1983, *Machine learning: An
     Artificial Intelligence Approach volume I*. Palo Alto, CA: Tioga.

19. Michalski, R., J. Carbonell, and T. Mitchell (eds.): 1986, *Machine learning: An
     Artificial Intelligence Approach Volume II*. Los Altos, CA: Morgan Kaufmann.

20. Ott, E.: 1993, *Chaos in Dynamical Systems*. New York, NY: Cambridge University
     Press.

21. Rabiner, L.: 1989, 'A tutorial on hidden Markov models and selected applications
     in speech recognition'. *Proceedings of the IEEE* **77**, 257–286.

22. Schalkoff, R.: 1992, *Pattern Recognition: Statistical, Structural and Neural Ap-
     proaches*. New York, NY: John Wiley & Sons.

23. Shil'nikov, L., A. Shil'nikov, D. Turaev, and L. Chua: 2000, *Methods of Qualitative
     Theory in Nonlinear Dynamics: Part I*. Singapore: World Scientific.

24. Shil'nikov, L., A. Shil'nikov, D. Turaev, and L. Chua: 2001, *Methods of Qualitative
     Theory in Nonlinear Dynamics: Part II*. Singapore: World Scientific.

25. Vapnik, V.: 1995, *The Nature of Statistical Learning Theory*. New York, NY:
     Springer-Verlag.

26. Weiss, S.: 1991, *Computer Systems that Learn? Classification and Prediction Meth-
     ods from Statistics, Neural Nets, Machine Learning and Expert Systems*. San Mateo,
     CA: Morgan Kaufmann.

BASIC PRINCIPLES OF DIRECT CHAOTIC COMMUNICATIONS

A. S. DMITRIEV[1]*, M. HASLER[2]†, A. I. PANAS[1] and
K. V. ZAKHARCHENKO[1]
[1]*Institute of Radioengineering and Electronics,
Russian Academy of Sciences,
Mokhovaya St., 11, Moscow, Russia*
[2]*Ecole Polytechnique Federale de Lausanne, Switzerland,*

Abstract

Basics of the theory of direct chaotic communications is presented. We introduce the notion of chaotic radio pulse and consider signal structures and modulation methods applicable in direct chaotic schemes. Signal processing in noncoherent and coherent receivers is discussed. The efficiency of direct chaotic communications is investigated by means of numerical simulation. Potential application areas are analyzed, including multiple access systems.

1. Introduction

Direct chaotic communication (DCC) systems are systems in which the information-carrying chaotic signal is generated directly in RF or microwave band [1–9]. Information is put into the chaotic signal by means of modulating either the chaotic source parameters or the chaotic signal after it is generated by the source. Consequently, information is retrieved from the chaotic signal without intermediate heterodyning.

The idea of direct chaotic systems and results of experiments with a wideband communication system operating in 900–1000 MHz band and providing transmission rates 10 to 100 Mbps were presented in Refs. [2–6]. The results of experiments with ultra-wideband direct chaotic circuit operating in 500–3500 MHz band with up to 200 Mbps rate are given in

---

* chaos@mail.cplire.ru
† martin.hasler@epfl.ch

A. Pikovsky and Y. Maistrenko (eds.), Synchronization: Theory and Application, 41–63.

Refs. [6–9]. The limit transmission rate in that system is estimated as 500–1000 Mbps. These experiments verified practical applicability of DCC and estimates of its performance.

A key notion of direct chaotic systems is the notion of chaotic radio pulse, which is a signal fragment whose duration is longer than the quasiperiod of chaotic oscillations. The frequency bandwidth of the chaotic radio pulse is determined by the bandwidth of the original chaotic source signal and is independent of the pulse duration in a wide range of duration variation. This makes the chaotic radio pulse essentially different of the classical radio pulse filled with a fragment of periodic carrier, whose frequency bandwidth Δf is determined by its duration τ

$$\Delta f \sim \frac{1}{\tau} \tag{1}$$

In this paper, the basics of the theory of direct chaotic transmission of information are given.

The paper layout is as follows.

In the first section, the scheme of direct chaotic communications is described. In the second, we consider the signal structure and modulation methods. In the third section, receiver models are described and their effectiveness is discussed. The system performance with noncoherent and coherent receivers is investigated in the fourth section. Then, we analyze in brief organization of multiple access, electromagnetic compatibility and ecological aspects.

2. Scheme of information transmission using chaotic radio pulses

Three main ideas constitute the basis of direct chaotic communication circuits [2–9]: (1) chaotic source generates oscillations directly in the prescribed microwave band; (2) information is put into the chaotic signal by means of forming the corresponding sequence of chaotic radio pulses; (3) information is retrieved from the microwave signal without intermediate heterodyning.

Block diagram of direct chaotic communication system in the cases of external and internal modulation is shown in Fig. 1.

The transmitter of the system is composed of a unit of oscillator control; a chaotic source that generates the signal directly in the frequency band of information transmission, i.e., in RF or microwave band; a keying-type modulator; an amplifier; an antenna; an information source; a message source encoder, and a channel encoder.

Chaotic source provides generation of the signal with the frequency bandwidth

$$\Delta F = F_u - F_l \tag{2}$$

where F_l and F_u are the lower and upper boundaries of the chaotic oscillation band. The chaotic signal frequency bandwidth is the frequency range, at which boundaries the power spectral density is -20 dB of the maximum within the range.

The central frequency

$$F_0 = (F_u + F_l)/2 \tag{3}$$

and the bandwidth ΔF of the generated signal may be adjusted by control unit. Modulator forms chaotic radio pulses and intervals between them.

Information that comes from an information source is transformed by the message source encoder into a signal that is fed to the channel encoder, which in turn transforms it into a modulating signal that controls the modulator. Modulator forms chaotic radio pulses either by means of multiplying chaotic signal and modulating pulses (the case of external modulation, Fig. 1, a), or by means of modulating the oscillator parameters (the case of internal modulation, Fig. 1, b). The duration of the formed chaotic pulses may be varied in the range $\tau \sim 1/\Delta F$ to $\tau \to \infty$.

The formed signal is put through amplifier and is emitted to free space with wideband antenna. Information stream can be formed by means of changing the intervals between the pulses, the pulse duration, the mean square amplitude of the pulses, or by means of combining these parameters.

For example, the stream can be formed so as to have constant rate of pulse positions and fixed duration of the pulses. In this case, the presence of a pulse at a certain prescribed position in the stream corresponds to transmission of symbol "1", and the absence of the pulse in the stream corresponds to the transmission of symbol "0".

The receiver (Fig. 1, c) is composed of a broadband antenna, a filter that passes the signal within the frequency band of the transmitter, a low-noise amplifier, and a signal processing system. The sequence of chaotic radio pulses comes to antenna and passes through filter and amplifier. The signal-processing system finds the pulses and determines their parameters and location in time domain. Then, the signal-processing system retrieves useful information from the signal either by means of integrating the pulse power over the pulse interval (noncoherent receiver), or by means of convolving the chaotic radio pulses with corresponding reference pulses generated in the receiver (coherent receiver).

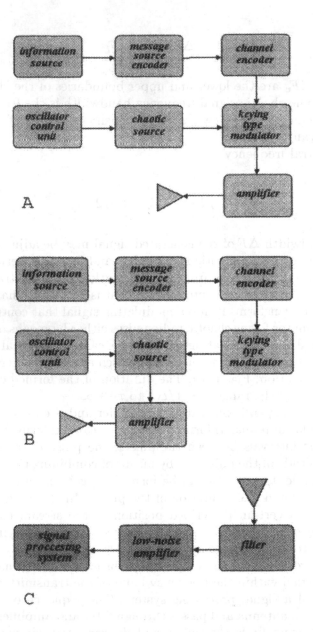

Figure 1. Block-diagram of direct chaotic communications system: (a) transmitter with external modulation; (b) transmitter with internal modulation; (c) receiver.

3. The signal structure and modulation methods

Information is transmitted by means of forming a sequence of chaotic radio pulses. Here, each pulse duration is τ, and the the pulse position is located within the time window of length T (in the average). The parameter

$$D = \frac{\tau}{T} \qquad (4)$$

will be called the duty cycle.

Let P be the power of the original chaotic signal, and let the transmission of information bit "1" be encoded by the presence of a chaotic radio pulse in the corresponding position and by "0" by its absence in this position. Then, the average power of the chaotic signal in the communication channel is equal to

$$P_{av} = \frac{P}{2}D. \qquad (5)$$

The factor 1/2 come from the fact that in average only half the pulses is present, provided that "1"'s and "0"'s have the same probability.

Let us denote the power spectral density of the chaotic signal by $s(f)$. As a rule, the spectral density is not constant within the frequency band ΔF. So, it is useful to know the mean-over-the-band spectral density of the signal

$$< s > = \frac{1}{F_u - F_l} \int_{Fl}^{F_u} s(f)\, df = \frac{1}{\Delta F} \int_{Fl}^{F_U} s(f)\, df. \qquad (6)$$

The signal base is a quantity [10–11]

$$B = 2\Delta F\tau \qquad (7)$$

Judging from the value of base B, elementary signals with base

$$B = 2\Delta F\tau \sim 1 \qquad (8)$$

and complex signals with

$$B = 2\Delta F\tau \gg 1 \qquad (9)$$

are distinguished.

Since $2\tau\Delta F$ is the signal base, then increasing the duration of chaotic radio pulse gives greater signal base. If the chaotic radio pulse duration is

$$\tau \gg \frac{1}{2\Delta F}$$

then the power spectrum of the sequence of chaotic radio pulses is practically the same as that of the original chaotic signal.

As an example, let us consider chaotic RF pulses obtained from the chaotic signal of a ring-structure oscillator with 2.5 freedom degrees that is passed through a band-pass filter (2000-5000 MHz) with −40 dD side-band suppression. Normalized equations of the oscillator are as follows:

$$T\dot{x} + x = F(z)$$
$$\ddot{y} + \alpha_1 \dot{y} + y = x$$
$$\ddot{z} + \alpha \dot{z} + \omega^2 z = \alpha_2 \dot{y}$$
$$F(z) = M \left[|z + E_1| - |z - E_1| + \tfrac{1}{2} \left(|z - E_2| - |z + E_2| \right) \right]$$

with the parameters set at $\alpha_1 = 0.0577; T = 0.7996; \alpha_2 = 0.2803; \omega = 0.7253; E_1 = 0.5; E_2 = 1; M = 20$.

The power spectrum of a periodic sequence of chaotic RF pulses as a function of the pulse length is presented in Fig. 2,a. As can be seen, in a wide range of the pulse length variation the form of the main spectrum lobe changes weakly and with a decrease of the pulse length the level of spectrum density outside the main band increases.

Dependencies for random sequences of chaotic pulses (Fig. 2, b) are presented in Fig. 3. As can be seen from comparison of Figs. 2, a and 2, b, in both cases the levels of the background spectrum outside the main lobe are approximately equal, however it is more smooth in the case of random sequence.

To understand the main features of chaotic radio pulse as an information carrier, let us compare it with two other carriers: harmonic signal and ultra-short ultra-wideband video pulses.

Radio pulses obtained by means of multiplication of a harmonic signal of frequency f_0 and video pulses of duration τ are elementary, because the bandwidth of such pulses is $\Delta F \sim 1/\tau$ and their base equals $B = \tau \Delta F \sim 1$ [13–14]. Simple-form ultra-short pulses [15], despite their super-wide bandwidth are also elementary, because the product of their duration by their bandwidth is also $B = 2\tau \Delta F \sim 1$. In contemporary communication systems, especially those operating under difficult signal propagation conditions (cellular systems, local wireless communications, etc.) large-base signals are preferably utilized. When operating with the signals based on harmonic carrier, spectrum spreading techniques (direct spread sequences or frequency hopping) are used, where the signal base is increased in proportion to the spectrum spreading factor [12–14]. To obtain large-base signals using ultra-wideband pulses, either the energy of several pulses is accumulated, or the pulse shape is complicated so as to increase their duration and retain the ultra-wide bandwidth.

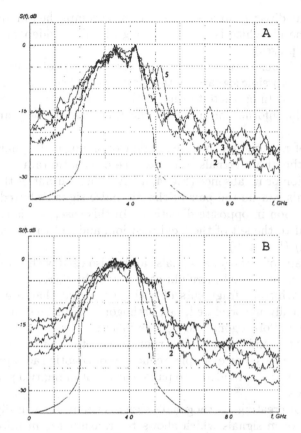

Figure 2. The power spectrum of the periodic (a) and random (b) sequence of chaotic RF pulses for four values of the pulse length τ: 1 – chaotic signal, 2 – τ_1=2ns, 3 – τ_2=1ns, 4 – τ_3=0.3ns, 5 – τ_4=0.25ns.

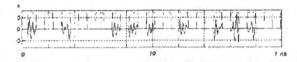

Figure 3. Waveform fragments of random sequence of chaotic RF pulses for the pulse length τ_3=0.3ns.

In contrast to the above signals, the signal base in DCC is determined by only the duration of chaotic radio pulses. No supplementary elements are necessary in the system to change the value of B. Moreover, at various transmission rates the input circuits of the receiver and the base values may be kept constant.

Note also that direct chaotic signals may be accomplished in any necessary frequency band (which is impossible, e.g., in ultra-wideband systems with ultra short pulses).

We distinguish below two classes of receivers. In a coherent receiver exact copies of all possible signals the transmitter could sent in a given time interval are present. In opposite to this, in the case of noncoherent receiver we do not use the information about of the waveform of the transmitted signal.

In the case of noncoherent receiver, a pair of orthogonal signals can be represented by the above signals: the presence of chaotic radio pulse in a prescribed position or its absence (Fig. 4, a). Another variant of the pair of orthogonal signals can be the pair of chaotic pulse signals shifted against some known position in opposite directions. In this case, "0" is related to the signal shifted to the left of the fixed position, and "1" to the right of it (in time domain) (Fig. 4, b).

Both these pairs of orthogonal signals can be also used with the coherent receiver. However, additional possibilities are available here.

Indeed, two arbitrary fragments of a chaotic signal that are of same length are practically uncorrelated, i.e., orthogonal. This fact can be used to organize pairs of "orthogonal" signals as follows.

The transmitter and receiver incorporate chaotic sources that form identical chaotic signals. For example, sequences of chaotic radio pulses are formed. Each sequence is divided into pairs, each containing two practically orthogonal signals.

Such a way of creating "orthogonal" signal system is naturally generalized to groups of m signals which allows to transmit $\log_2 m$ information bits during one active interval.

With coherent receiver antipodal signals can also be used. As in the case of orthogonal signals, identical sequences of chaotic radio pulses are formed in the transmitter and receiver. When transmitting "0", the radio pulse is inverted by sign, while by transmission of "1" it is emitted in its original form.

Geometrically, the discussed signal systems can be presented as in Fig. 5.

Note that uncorrelated character (orthogonality) of chaotic radio pulses is approximate. The same is the situation with the chaotic radio pulse energy. In general, it varies from pulse to pulse and this fact must be taken into account by calculation of communication system parameters.

4. Noncoherent and coherent receiver models

Information transmitted with chaotic radio pulses can be received with two main methods: noncoherently or coherently. Here we restrict our analysis of

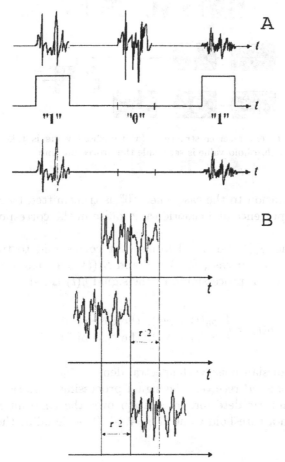

Figure 4. Examples of orthogonal pairs of signals for direct chaotic communication systems: (a) simplest system of signals; (b) pulse position modulation.

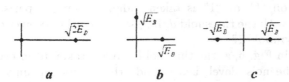

Figure 5. Geometrical interpretation for orthogonal (a, b) and antipodal (c) signals.

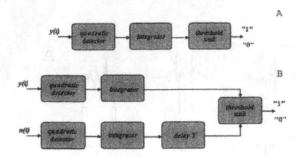

Figure 6.   Noncoherent receiver structure: (a) threshold value is taken outside the processing unit; (b) threshold value is set inside the processing unit.

the receiver operation to the case when "0" is transmitted by the absence and "1" by the presence of a chaotic radio pulse in the corresponding time position.

Let $S_0(t)$ and $S_1(t)$ be signal forms that correspond to two different transmitted bits. In our case, $S_0(t) = 0$ and $S_1(t)$ is a chaotic radio pulse. At the input of signal-processing unit the signal $y(t)$ is fed

$$y(t) = \begin{cases} S_0(t) + \eta(t), & \text{the case of "0"} \\ S_1(t) + \eta(t), & \text{the case of "1"} \end{cases} . \tag{10}$$

where $\eta(t)$ is Gaussian noise with spectral density N_0.

In the *noncoherent* receiver, the signal processing comprises its detection with a quadratic detector, integration over the time interval τ, and comparison with a threshold value (Fig. 4, a). The signal at the threshold unit is

$$d_0 = \int_0^T \eta^2(t) dt,$$

if the transmitted signal is "0", and

$$d_1 = \int_0^T (S(t) + \eta(t))^2 dt,$$

is symbol "1" is transmitted.

The decision on "0" or "1" is taken judging from comparison of the obtained value d_i with the threshold d. If $d_i < d$, then the receiver concludes "0", otherwise, symbol "1".

In the circuit in Fig. 6, a, the threshold value is taken from preliminary observations of the noise level, i.e., outside the processing unit. Alternatively, in the circuit presented in Fig. 6,b, the noise energy is estimated on the time interval τ preceding the position with the chaotic radio pulse. With this estimate, the threshold value is set automatically.

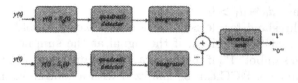

Figure 7.   Coherent receiver structure.

In the *coherent* receiver, the signal forms $S_0(t)$ and $S_1(t)$ that correspond to two different transmitted signals are assumed to be present. In our case, $S_0(t) = 0$ and $S_1(t)$ is a chaotic radio pulse.

Then the following operations are performed in the signal-processing unit (Fig. 7).

The incoming signal $y(t)$ is fed to the upper and lower processing channels, where $S_0(t) \equiv 0$ and $S_1(t)$ are subtracted from it, respectively. The obtained difference signals are detected with a quadratic detector and integrated over the time interval τ, equal to the chaotic radio pulse duration. The processed signals are directed to a summer, and after that the threshold unit makes decision which symbol is received.

Let a signal corresponding to symbol "0" come to the receiver. Then, in the upper channel we have

$$d_0 = \int\limits_0^T \eta^2(t)dt,$$

while in the lower chanel

$$d_1 = \int\limits_0^T (S_1(t) - \eta(t))^2 dt$$

and the difference $d_0 - d_1 < 0$.

If the signal corresponding to "1" comes to the receiver, then in the upper channel we have

$$d_0 = \int\limits_0^T (S_1(t) + \eta(t))^2 dt,$$

in the lower chanel

$$d_1 = \int\limits_0^T \eta^2(t)dt$$

and the difference $d_0 - d_1 > 0$.

Thus, the threshold for decision is zero and the threshold unit makes decision on symbol "0" received, if the signal at the summer's output is negative, and on symbol "1", if the signal at the summer's output is positive.

Noise resistance of DCC can be analyzed with a continuous model discussed above, as well as with the signal sampled in time domain. In the latter case, signals samples $S_0(t)$ and $S_1(t)$ are substituted by signals $S_0(i)$ and $S_1(i)$, and noise $\eta(t)$ by noise samples $\eta(i)$, where $S_0(i) = S_0(i\tau/n)$, $S_1(i) = S_1(i\tau/n)$, $\eta(i) = \eta(i\tau/n)$.

The filter passband is assumed to be matched with the chaotic signal band. So, the noise signal has the same correlation time and bandwidth as the chaotic signal.

In discrete case, main relations for noncoherent and coherent receivers are as follows.

Noncoherent case.

$d_0 = \sum_{i=1}^{n} \eta^2(i)-$ for symbol "0" transmitted.

$d_1 = \sum_{i=1}^{n} (S_1(i) + \eta(i))^2-$ for symbol "1".

The problem of noncoherent receiver in discrete case as in continuous case is that d_0 and d_1 are compared not to each other, but each with a threshold that is estimated separately, which contributes additional uncertainty.

Coherent case. At the output of the upper channel we have

$d_0 = \sum_{i=1}^{n} \eta^2(i)-$ for symbol "0" transmitted and

$d_0 = \sum_{i=1}^{n} (S_1(i) + \eta(i))^2-$ for symbol "1".

At the output of the lower channel we have:

$d_1 = \sum_{i=1}^{n} (S_1(i) - \eta(i))^2-$ for symbol "0" transmitted and

$d_1 = \sum_{i=1}^{n} \eta^2(i)-$ for symbol "1".

As in continuous case, if the signal at the summer's output is negative, the decision is symbol "0" received, if positive, then the received symbol "1".

5. DCC performance

We expect that the error probability for the discussed modulation and reception methods approximately correspond to standard characteristics of signal systems that permit geometrical interpretation as in Fig. 5, namely (for error probability $P = 10^{-3}$, Gaussian noise spectral density N_0 and average energy of chaotic radio pulse per transmitted information bit E_b): noncoherent receiver, orthogonal signals (Fig. 5, a) – $E_b/N_0 = 13$–14 dB; coherent receiver, orthogonal signals (Fig. 5, a) – $E_b/N_0 = 10$–12 dB; coherent receiver, orthogonal signals (Fig. 5, b) – $E_b/N_0 = 10$–12 dB; coherent receiver, antipodal signals (Fig. 5, c) – $E_b/N_0 = 7$–8 dB.

We test this conjecture by means of direct numerical simulation. Note that the signal energy varies from pulse to pulse due to its chaotic nature which is not the case for classical career signal. Besides, in the noncoherent receiver, one expects a decrease of the receiver efficiency at large values of the signal base. The results of direct simulation for a pair of orthogonal signals (Fig. 5, a) are given in Figs. 8–11. By the simulation, the chaotic signals were assumed to have three variants of instantaneous chaotic signal distribution: (1) Gaussian distribution, (2) uniform distribution with zero mean, and (3) ±1 values with equal probability.

As a whole, Fig. 8 (noncoherent receiver) and Fig. 9 (coherent receiver) confirm the preliminary estimates. However, statistical distribution of the signal values appears to be important. For example, for the *noncoherent receiver* the worst is the Gaussian distribution (Fig. 8, a). In this case, error probability $P \leq 10^{-3}$ is achieved only beginning from the value $(E_b/N_0) > 15$ dB in the range of $20 < B < 60$. For small value of B error probability $P \leq 10^{-3}$ is never achieved at any reasonable values of E_b/N_0. This is associated with long tails of the distribution of the sum of squares of normally distributed random variable.

For uniform distribution (Fig. 8, b) the situation is much better. In this case, error probability $P \leq 10^{-3}$ is achieved beginning from the value $(E_b/N_0) > 14$ dB in the range of $10 < B < 30$. For small $B \sim 4$–6 error probability $P \leq 10^{-3}$ is also achievable, though at a slightly greater value of E_b/N_0, namely, at value $(E_b/N_0) > 15$ dB.

Finally, for the distribution of a kind of random telegraph signal the most effective are the signals with small base values. In this case, error probability $P \leq 10^{-3}$ can be obtained already at the value $(E_b/N_0) > 11$ dB.

Note that the receiver efficiency decreases with increasing base beginning from the base $B > 50$ and at the base $B = 300$ error probability $P \leq 10^{-3}$ is obtained in all three cases only for $(E_b/N_0) > 16.5$ dB.

In the case of the *coherent receiver* (Fig. 9), asymptotically at $B \to \infty$ for all distributions error probability $P \leq 10^{-3}$ is achieved if $(E_b/N_0) > 10$ dB. However, in the case of random telegraph distribution the asymptotic condition is satisfied already at low values of B, whereas in the worst case (Gaussian distribution) it is achieved only for $B > 100$. Note also that in the case of Gaussian distribution and small base, $P \leq 10^{-3}$ cannot be obtained by means of moderate increase of E_b/N_0 (Fig. 9, a). At the same time, this can be easily achieved with the uniform-distribution signal.

As follows from the above analysis, Gaussian-distribution signals are the least favorable for direct chaotic communications as in the case of noncoherent as well as coherent receivers. Under all other equal conditions they provide transmission rates 3–4 times lower than the signals with uniform or random telegraph distribution. In the range where the transmission rate for all three types of signals is approximately equal ($10 < B < 50$), the systems with Gaussian-distribution signals provide $P \leq 10^{-3}$ at (E_b/N_0) about 1–1.5 dB greater than the systems with two other types of signals.

As follows from the above, in particular, in the case of uniform or random telegraph distributions the limit information-carrying capacity of the channel can reach its bandwidth (e.g., with 1 GHz bandwidth the transmission rate can achieve 1 Gbit/sec), whereas in the case of Gaussian signal the limit rate is 3–4 times less.

It is interesting to note that for the signal with random telegraph distribution the efficiency of noncoherent and coherent receivers at low base values is nearly the same.

Finally, let us compare the effectiveness of signal energy accumulation with increasing base for noncoherent and coherent receivers. If in the range of base values $10 < B < 50$ the coherent receiver is 2.5–4.5 dB more effective than the noncoherent one, then for base 200 the difference in efficiency is already 6 dB (Figs. 8 and 9). Error probabilities are presented for various average signal-to-noise ratio average values in Fig. 10 (noncoherent receiver) and in Fig. 11 (coherent receiver). All curves are for the case of duty cycle $D = 1/2$. As can be seen in the Figures, at $P < 10^{-3}$ both with coherent receiver (Fig. 11) and with noncoherent receiver (Fig. 10) the average power of the transmitted signal can be lower than the noise power. This effect is strengthened in proportion to the duty cycle.

Real chaotic signals have finite amplitude and their distributions do not have long tails as the Gaussian does. Therefore, the estimates pertaining to uniformly distributed signal and to random telegraph signal are more suitable for them. Moreover, the model with random telegraph signal matches especially good to the signals with phase chaos, that are obtained, e.g., in phase-locked loop systems [16].

6. Multiple access

All three conventional techniques of sharing the channel between many users, such as frequency-division multiple access (FDMA), time-division multiple access (TDMA) and code-division multiple access (CDMA), can be accomplished in DCC.

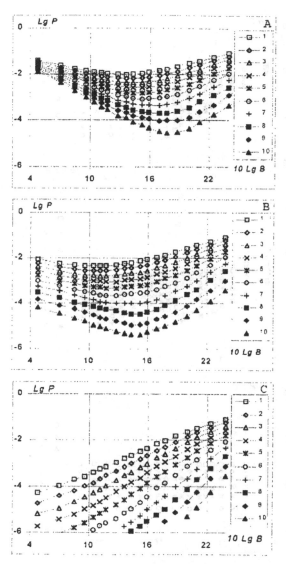

Figure 8. Noncoherent receiver efficiency as a function of signal base (B) for different E_b/N_0 values: (a) chaotic signal with Gaussian distribution; (b) chaotic signal with uniform distribution; (c) chaotic signal with equal probability of ±1 values. 1 – 12.5 dB (contour square), 2 – 13 dB (contour rhombus), 3 – 13.5 dB (contour triangle), 4 – 14 dB (diagonal cross), 5 – 14.5 dB (asterisk), 6 – 15 dB (contour circle), 7 – 15.5 dB (vertical cross), 8 – 16 dB (solid square), 9 – 16.5 dB (solid rhombus), 10 – 17 dB (solid triangle).

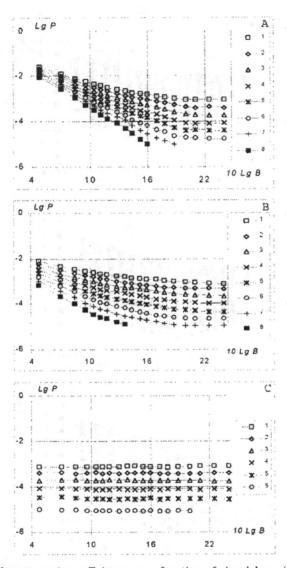

Figure 9.   Coherent receiver efficiency as a function of signal base (B) for different E_b/N_0 values: (a) chaotic signal with Gaussian distribution; (b) chaotic signal with uniform distribution; (c) chaotic signal with equal probability of ±1 values. 1 – 10 dB (contour square), 2 – 10.5 dB (contour rhombus), 3 – 11 dB (contour triangle), 4 – 11.5 dB (diagonal cross), 5 – 12 dB (asterisk), 6 – 12.5 dB (contour circle), 7 – 13 dB (vertical cross), 8 – 13.5 dB (solid square).

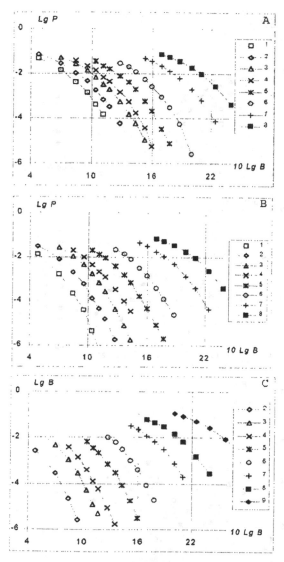

Figure 10. Noncoherent receiver. Error probability as a function of signal base (*B*) for different SNR values: (a) chaotic signal with Gaussian distribution; (b) chaotic signal with uniform distribution; (c) chaotic signal with equal probability of ±1 values. 1 – 7 dB (contour square), 2 – 5 dB (contour rhombus), 3 – 3 dB (contour triangle), 4 – 2 dB (diagonal cross), 5 – 0 dB (asterisk), 6 – -2 dB (contour circle), 7 – -5 dB (vertical cross), 8 – -7 dB (solid square), 9 – -10 dB (solid rhombus).

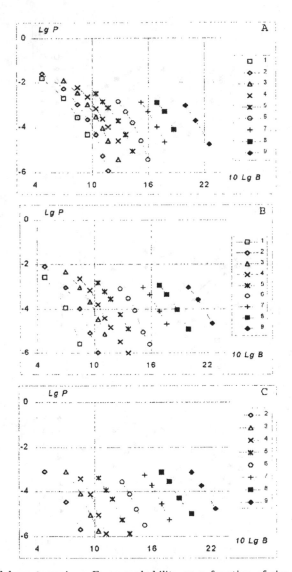

Figure 11.   Coherent receiver. Error probability as a function of signal base (B) for different SNR values: (a) chaotic signal with Gaussian distribution; (b) chaotic signal with uniform distribution; (c) chaotic signal with equal probability of ±1 values. $1 - 7$ dB (contour square), $2 - 5$ dB (contour rhombus), $3 - 3$ dB (contour triangle), $4 - 2$ dB (diagonal cross), $5 - 0$ dB (asterisk), $6 - -2$ dB (contour circle), $7 - -5$ dB (vertical cross), $8 - -7$ dB (solid square), $9 - -10$ dB (solid rhombus).

The most interesting is time-division multiple access discussed below.

Time-division multiple access. Let there be an ideal Gaussian channel with bandwidth W and the noise spectrum density N_0. The throughput of such a channel in the case of an information signal with mean power P equals to

$$C = W \log_2(1 + \frac{P}{W N_0}). \tag{11}$$

In the case of DCC, $W = \Delta F$, where ΔF is the chaotic signal bandwidth. Let us compare the throughput of this channel with that of the channel with K TDMA users in assumption that the mean signal power of ith user is $P_i = P$ ($i = 1..K$).

In TDMA system each user transmits information during $1/K$th part of the total time in the entire frequency bandwidth ΔF with the signal power KP. So, the throughput per each user is

$$C_K = \frac{\Delta F}{K} \log_2(1 + \frac{KP}{W N_0}). \tag{12}$$

The total throughput of a multiple-access DCC channel is

$$C_K = \Delta F \log_2(1 + \frac{KP}{W N_0}), \tag{13}$$

i.e., theoretically it is somewhat higher than the capacity of a channel with a single user. It is provided due to an increase of the signal-to-noise ratio (factor K is under logarithm).

Relations (8)–(10) show that multiple access can be efficiently realized in DCC by means of time division.

Let $E_b = P/C_K$ be the signal energy per bit of transmitted information. Then, relation (9) can be rewritten as

$$\frac{K C_K}{\Delta F} = \log_2(1 + \frac{K C_K}{\Delta F} \frac{E}{N_0}), \tag{14}$$

and (8) as

$$\frac{C}{\Delta F} = \log_2(1 + \frac{C}{\Delta F} \frac{E}{N_0}). \tag{15}$$

All relations (8)–(12) are limits. In real schemes, the throughput in multiple-access mode is determined by the single-user system throughput as a function of the signal-to-noise ratio. For example, in TDMA DCC the same throughput per user can be achieved both with coherent or noncoherent receivers. However, with noncoherent receiver it can be achieved at a slightly higher value of E_b/N_0.

Multiple access in DCC can also be organized using packet data transmission under IEEE 802.11 standard.

7. Electromagnetic compatibility

In the case of small duty cycle, the sequence of wideband (ultra-wideband) chaotic radio pulses interferes with the signals of conventional radio circuits only on very small time intervals. For example, at a rate of 10^6 bps and the length of each chaotic pulse 10^{-8} sec., the duty cycle is $D = 10^{-2}$, so, the time of interference is 0.5% of the system operation time. Besides, the average emitted power is by a factor of 200 lower than the mean power during emission of chaotic radio pulses. If, e.g., the transmitter power during the pulse emission is 200 mW, then the average power is only 1 mW.

Let us consider the interaction of narrowband and wideband signals in more detail. A typical interference of narrowband communication systems is wideband noise. The method combating the noise is frequency filtering. By means of matching the receiver bandwidth with that of the received signal, one manages to cut the most part of the noise energy off. The remaining noise energy $\frac{N_0}{2}\Delta F$ makes the energy of interference.

With wideband and ultra-wideband DCC systems the situation is different. Here, besides the Gaussian white noise the interference is represented by narrowband signals of the devices operating within its frequency range. In noncoherent receiver, the average power of the interference signal at the input is

$$P_{int} = \int\limits_{F}^{F} \frac{N_0}{2} df + \int\limits_{F_1}^{F} S(f) df, \tag{16}$$

where $S_{nb\eta}(f)$ is the spectral density of narrowband signal. Take for simplicity that there is a single narrowband signal whose spectral density is constant within certain frequency range and zero outside that range.

Let us denote the bandwidth ratio of the chaotic signal to the narrowband signal by M, and the power ratio of the narrowband signal to the Gaussian signal within the bandwidth of the narrowband interference signal, necessary to receive information with admissible error probability, by L. (This ratio is equal to the ratio of spectral densities). Then the total power of the interference signal is

$$P_n = (1 + \frac{L}{M})\frac{N_0}{2}\Delta F. \tag{17}$$

Example. The receiver bandwidth is 100 MHz, the narrowband signal bandwidth is 1 MHz, $L = 15$ dB, then

$$P_{int} = (1 + \frac{30}{100})P_{int}^0 = 1.3P_{int}^0, \tag{18}$$

where P_{int}^0 is the power of interference determined by Gaussian noise in the frequency range of the chaotic signal.

If the information sequence of chaotic radio pulses has duty cycle D, then the effective average interference power is lower by a factor of D

$$P_e = D \cdot P_{int}. \tag{19}$$

Let the same volume of information be transmitted with the rate 1 Mbps using narrowband and wideband signals, and the chaotic signal duty cycle be 10^{-2} and $L = (E_b/N_0) = 15$ dB. Denote the energy of the narrowband signal, obtained by wideband receiver when receiving one information bit, by $E_{nb,n}$. As follows from (15) and (16), effective excess to the Gaussian noise level is 30%, and interference added by the narrowband signal is $(E_{nb,n}/E_b) = -20$ dB. Thus, distortions induced by narrowband information signal are below the admissible level of $L = 15$ dB, consequently, are not an interference for the wideband information signal.

Consider now the effect of the wideband information signal on the narrowband information signal under the same conditions $((E_b/N_0) = 15$ dB). Denote the energy of the wideband chaotic signal, obtained by narrowband receiver when receiving one information bit, by $E_{w,n}$. With the duty cycle $D = 10^{-2}$ taken into account, this energy is 5 dB less than the interference energy induced by Gaussian noise. Hence, $(E_{w,n}/E_b) = -20$ dB, i.e., narrowband signal distortions due to reception of a part of wideband signal are also negligible.

Thus, information-carrying traditional narrowband and wideband (ultrawideband) direct chaotic signals have but a little effect on each other and in many cases can be used together. Actually, this gives a chance of repeated use of the already occupied regions of microwave band.

8. Ecological safety

The degree of an effect of electromagnetic radiation on living systems is determined by the total power of electromagnetic radiation and by its structure. In the case of signals based on harmonic carrier, distinct spectral components are present in the electromagnetic spectrum that can have selective (resonant) effect on various subsystems of living organisms. In the case of ultra-short almost periodic video pulses, potential danger may be in "percussive" periodic effect of electromagnetic pulses.

Unstructured in time domain and "spread" over the frequency band, direct chaotic signals are less dangerous, because their potential negative effect is determined by only an increase of the environmental electromagnetic background radiation. Besides, in the majority of practically interesting cases the added radiation level is below the natural background. Thus, DCC are ecologically more safe than traditional radio systems.

9. Conclusions

In conclusion, note two other useful aspects of DCC. The first is the stability of communications in multipath environment.

Multipath propagation is a less problem to DCC than to conventional communication systems. Actually, Rayleigh fading that takes place due to signal interference by multipath propagation is caused by the narrow-band character of signals. In order to make interference between individual chaotic radio pulses possible, certain conditions must be fulfilled. However, even if such an interference takes place, it doesn't lead to as unpleasant consequences as in the case of sinusoidal signals, because wideband chaotic signals have rapidly decreasing autocorrelation functions.

The second aspect is simplicity of the transmitter and receiver design and, as a consequence, low cost of mass production. The transmitter is a chaotic source containing a little number of components, and information is put directly to the carrying chaotic signal. So, the transmitter layout is rid of a number of elements used in conventional communication systems, and restrictions to the remaining parts are more soft than in classical systems. In particular, there are no strong restrictions on the linearity of the output amplifier, which decreases the cost and energy consumption.

The receiver is also more simple in structure than the narrowband receiver, because the stage of signal processing at an intermediate (heterodyne) frequency is not necessary here. Besides, in contrast to conventional spread-spectrum receivers, the control circuits operate here not at microwave frequencies but at pulse repetition frequencies. Preliminary analysis shows that DCC transceivers can be accomplished in the form of single inexpensive chip.

Acknowledgments

We are grateful to Yu.V. Andreev, L.V. Kuzmin and S.O. Starkov for useful discussions. The work was supported in part by the Russian Foundation for Basic Research, project no. 02-02-16802 and the Swiss National Science Foundation, SOPES project no. 75SUPJ062310.

References

1. A.S. Dmitriev, B.Ye. Kyarginsky, N.A. Maximov, A.I. Panas, S.O. Starkov. Prospects of Direct Chaotic Communications Design in RF and Microwave Band. Radiotehnika. **42**, no. 3, 2000 9-20.

2. A.S. Dmitriev, B.Ye. Kyarginsky, N.A. Maximov, A.I. Panas, S.O. Starkov. Direct Chaotic Communications in Microwave Band. Preprint IRE RAS, no. 1(625), Moscow, 2000.

3. A.S. Dmitriev, B.Ye. Kyarginsky, A.I. Panas, S.O. Starkov. Direct Chaotic Communication Schemes in Microwave Band. Radiotehnika I Elektronika. **46**, no. 2, 2001 224-233.

4. A. Dmitriev, B. Kyarginsky, A. Panas, S. Starkov. Direct Chaotic Communication System. Experiments. Proc. of NDES'01. Delft. Netherlands, 2001 157-160.

5. A.S. Dmitriev, A.I. Panas, S.O. Starkov. Direct Chaotic Communication in Microwave Band. Electronic NonLinear Science Preprint, nlin.CD/0110047, 2001.

6. A. Dmitriev, B. Kyarginsky, A. Panas, D. Puzikov, S. Starkov. Experiments on Ultra Wideband Direct Chaotic Information Transmission in Microwave Band. Proc. of NDES'02. Turkey, 21-23 june 2002 5-1 – 5-4.

7. A.S. Dmitriev, A.I. Panas, D.Yu. Puzikov, S.O. Starkov. Wideband and Ultra-wideband Direct Chaotic Communications. Proc. of ICCSC'02, St. Petersburg, Russia, 26-28 June 2002 291-295.

8. A.S. Dmitriev, B.Ye. Kyarginsky, A.I. Panas, D.Yu. Puzikov, S.O. Starkov. Experiments on Ultrawide Band Direct Chaotic Information Transmission in Microwave Band. Radiotehnika I Elektronika. **47**, no. 10, 2002 (in print).

9. A.S. Dmitriev, B.Ye. Kyarginsky, A.I. Panas, S.O. Starkov. Experiments on Direct Chaotic Communications in Microwave Band. Int. J. of Bifurcation and Chaos. **17**, no. 8, 2003 (in print).

10. N.T. Petrovich, M.K. Razmahnin. *Communication Systems with Noise Like Signals.* (Sovetskoe Radio, Moscow, 1969).

11. L.Ye. Varakin. *Theory of complex Signals.* (Sovetskoe Radio, Moscow, 1970).

12. J.G. Proakis, M. Salehi. *Contemporary Communications System using Matlab.* (CA: Book/Cole, 2000).

13. J.G. Proakis. *Digital Communications.* (McGraw-Hill Inc, NY, 1995).

14. S. Haykin. *Communication Systems.* (John Wiley & Sons, Inc, NY, 1994).

15. M.Z. Win, R.A. Scholtz. Impulse radio: how it works? IEEE Comm. Lett. **2**, no. 1, 1998 10.

16. V.D. Shalfeev, G.V. Osipov, A.K. Kozlov, A.R. Voilkovsky. Chaotic Oscillations – Generation, Synchronization, Control. Uspehi Sovremennoy Radioelektroniki. no. 10, 1997 27-49.

PREVALENCE OF MILNOR ATTRACTORS AND CHAOTIC ITINERANCY IN 'HIGH'-DIMENSIONAL DYNAMICAL SYSTEMS

KUNIHIKO KANEKO

Department of Pure and Applied Sciences, College of Arts and Sciences, University of Tokyo, Komaba, Meguro-ku, Tokyo 153, Japan

Abstract

Dominance of Milnor attractors in high-dimensional dynamical systems is reviewed, with the use of globally coupled maps. From numerical simulations, the threshold number of degrees of freedom for such prevalence of Milnor attractors is suggested to be 5 ∼ 10, which is also estimated from an argument of combinatorial explosion of basin boundaries. Chaotic itinerancy is revisited from the viewpoint of Milnor attractors. Relevance to neural networks is discussed.

1. Introduction

High-dimensional dynamical systems often have many attractors. Indeed, such systems with multiple attractors have been discussed to be relevant to biological memory, information processing, and differentiation of cell types[1, 2]. There, to make connection with problems in biological networks, stability of attractors against external perturbation, switching among attractors by external operation, and spontaneous itinerancy over several lower dimensional states are important. In the present paper, we survey a universal aspect with regards to these points.

One of the simplest model for such high-dimensional dynamical systems is globally coupled dynamical systems. In particular, "globally coupled map" (GCM) consisting of chaotic elements [3] has been extensively studied, as a simple prototype model. A standard model for such GCM is given by

A. Pikovsky and Y. Maistrenko (eds.), Synchronization: Theory and Application, 65–77.
© 2003 *Kluwer Academic Publishers. Printed in the Netherlands.*

$$x_{n+1}(i) = (1 - \epsilon)f(x_n(i)) + \frac{\epsilon}{N}\sum_{j=1}^{N} f(x_n(j)) \tag{1}$$

where n is a discrete time step and i is the index of an element ($i = 1, 2, \cdots, N$ = system size), and $f(x) = 1 - ax^2$. The model is just a mean-field-theory-type extension of coupled map lattices (CML)[4].

Through the average interaction, elements are tended to oscillate synchronously, while chaotic instability leads to destruction of the coherence. When the former tendency wins, all elements oscillate coherently, while elements are completely desynchronized in the limit of strong chaotic instability. Between these cases, elements split into clusters in which they oscillate coherently. Here a cluster is defined as a set of elements in which $x(i) = x(j)$. Attractors in GCM are classified by the number of synchronized clusters k and the number of elements for each cluster N_i. Each attractor is coded by the clustering condition $[k, (N_1, N_2, \cdots, N_k)]$.

As has been studied extensively, the following phases appear successively with the increase of nonlinearity in the system (a in the above logistic map case) [3]:

(i) **Coherent phase**; (ii) **Ordered phase**; (iii) **Partially ordered phase**; (iv) **Desynchronized phase**:

In (i), a completely synchronized attractor ($k = 1$) exists, while all attractors consist of few ($k = o(N)$) clusters in (ii). In (iii), attractors with a variety of clusterings coexist, while most of them have many clusters ($k = O(N)$). Elements are completely desynchronized, i.e., $k = N$ for all (typically single) attractors in (iv). The above clustering behaviors have universally been confirmed in a variety of systems (see also [5]).

In the partially ordered (PO) phase, there are a variety of attractors with a different number of clusters, and a different way of partitions $[N_1, N_2, \cdots, N_k]$. In this phase, there are a variety of partitions as attractors. As an example, we measured the fluctuation of the partitions, using the probability Y that two elements fall on the same cluster. In the PO phase, this Y value varies by attractors, and furthermore, the variation remains finite even in the limit of $N \to \infty$ [6, 7]. In other words, there is no 'typical' attractor in the thermodynamic limit. This is similar with the 'non-self-averaging' in Sherrington-Kirkpatrick model in spin glass[8].

2. Attractor Strength and Milnor Attractors

In the partially ordered (PO) phase and also in some part of ordered phase, there coexist a variety of attractors depending on the partition. To study the stability of an attractor against perturbation, we introduce the return probability $P(\sigma)$, defined as follows[9, 10]: Take an orbit point $\{x(i)\}$ of an attractor in an N-dimensional phase space, and perturb the point to $x(i) + \frac{\sigma}{2}rnd_i$, where rnd_i is a random number taken from $[-1, 1]$, uncorrelated for all elements i. Check if this perturbed point returns to the original attractor via the original deterministic dynamics (1). By sampling over random perturbations and the time of the application of perturbation, the return probability $P(\sigma)$ is defined as (# of returns)/ (# of perturbation trials). As a simple index for robustness of an attractor, it is useful to define σ_c as the largest σ such that $P(\sigma) = 1$. This index measures what we call the *strength* of an attractor.

The strength σ_c gives a minimum distance between the orbit of an attractor and its basin boundary. Note that σ_c can be small, even if the basin volume is large, if the attractor is located near the basin boundary.

In contrast with our naive expectation from the concept of an attractor, we have often observed 'attractors' with $\sigma_c = 0$, i.e., $P(+0) \equiv \lim_{\delta \to 0} P(\delta) < 1$. If $\sigma_c = 0$ holds for a given state, it cannot be an "attractor" in the sense with asymptotic stability, since some tiny perturbations kick the orbit out of the "attractor". The attractors with $\sigma_c = 0$ are called Milnor attractors[11, 12]. In other words, Milnor attractor is defined as an attractor that is unstable by some perturbations of arbitrarily small size, but globally attracts orbital points. (Originally, Milnor proposed to include all states with the basin of attraction of a positive Lebesgue measure, into the definition of an attractor[11]. Accordingly, attractors by his definition include also the usual attractor with asymptotic stability. Here we call Milnor attractor, only if it does not belong to the latter. If this Milnor attractor is chaotic, the basin is considered to be riddled [13, 14]. This is the case for the present GCM model.) Since it is not asymptotically stable, one might, at first sight, think that it is rather special, and appears only at a critical point like the crisis in the logistic map[11]. To our surprise, the Milnor attractors are rather commonly observed around the border between the ordered and partially ordered phases in our GCM (see Fig.2). Attractors with $\sigma_c = 0$ often have a large basin volume, which sometimes occupy almost 100 % of the total phase space.

With regards to the basin volume of Milnor attractors, the change of the behavior of the GCM (1) with the increase of the parameter a is summarized as follows: *a few attractors with small numbers of clusters;* → *increase of the number of attractors with stable and Milnor attractors coexisting;* → *decrease of the number of attractors with some remaining Milnor attractors with large basin fractions;* → *only a single or a few stable attractors with complete de-synchronization remain. Milnor attractors no longer exist.*

Then, why is the basin volume of Milnor attractors so large for some parameter regimes? To answer the question, robustness of global attraction

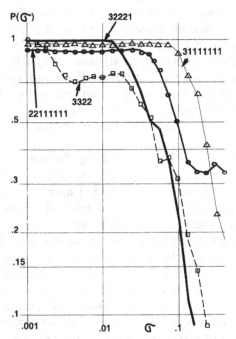

Figure 1. $P(\sigma)$ for 4 attractors for $a = 1.64$, and $N = 10$. 10000 initial conditions
are randomly chosen, to make samplings. $P(\sigma)$ is estimated by sampling over 1000
possible perturbations for each σ. Plotted are robust attractors [32221] ($\sigma_c \approx .01$), [3322]
($\sigma_c \approx .0012$), and Milnor attractors [31111111], [22111111]. The basin volume of the latter
two occupies 42% and 29% of the phase space.

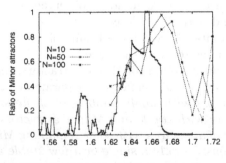

Figure 2.   The fraction of the basin ratio of Milnor attractors, plotted as a function of
a, for $N = 10, 50, 100$

is a key. Note there are a large number of attractors at the border between O and PO. Most of the attractors lose the stability around the parameter regime successively. When the stability of an attractor is lost, there appears a set of points in the vicinity of the attractor, that are kicked out of it through the temporal evolution, while the global attraction still remains. This is a reason why fragile attractors are dominant around the PO phase. In Fig.2, we have plotted the sum of basin volume rates for all the Milnor attractors. Dominance of Milnor (fragile) attractors is clearly seen.

Attractors are often near the crisis point and lose or gain the stability at many parameter values in the PO phase. Furthermore, the stability of an attractor often shows sensitive dependence on the parameter. It is interesting to see how $P(+0)$ and basin volume change with the parameter a, when an attractor loses asymptotic stability. As shown in [15], the basin volume of an attractor often has a peak when the it loses the stability and then decreases slowly as $P(+0)$ gets smaller than unity, where the attractor becomes a Milnor one. Although the local attraction gets weaker as $P(+0)$ is smaller than 1, the global attraction remains. It is also noted that if $P(+0)$ equals 1 or not often sensitively depends on the parameter a, while the basin volume shows smooth dependence on the parameter. The basin volume reflects on global attraction, while the $P(+0)$ depends on local structure in the phase space with regards to collision of an attractor and its basin boundary. This is the reason why the former has a smooth dependence of the parameter, in contrast to the latter.

Remark. Coexistence of attractors with different degrees of stability makes us expect that noise is relevant to the choice of the attractor the GCM settles to. One might then suspect that such Milnor attractors must be weak against noise. Indeed, by a very weak noise with the amplitude σ, an orbit at a Milnor attractor is kicked away, and if the orbit is reached to one of attractors with $\sigma_c > \sigma$, it never comes back to the Milnor attractor. In spite of this instability, however, an orbit kicked out from a Milnor attractor is often found to stay in the vicinity of it, under a relatively large noise[10]. The orbit comes back to the original Milnor attractor before it is kicked away to other attractors with $\sigma_c > \sigma$. Furthermore, by a larger noise, orbits sometimes are more attracted to Milnor attractors. Such attraction is possible, since Milnor attractors here have global attraction in the phase space, in spite of their local instability.

3. Magic Number 7 ± 2 in Dynamical Systems?

Milnor attractors can exist in low dimensional dynamical systems like a two-dimensional map as well. When changing the parameter of a dynamical

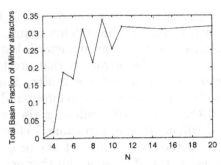

Figure 3.   The average fraction of the basin ratio of Milnor attractors. After the basin fraction of Milnor attractor is computed as in Fig.1, the average of the ratios for parameter values $a = 1.550, 1.552, 1.554, \cdots 1.72$ is taken. This average fraction is plotted as a function of N. (In this class of models, the fraction of Milnor attractors is larger for odd N than for even N. Note that two clusters with equal cluster numbers and anti-phase oscillations generally have less chaotic instability. A globally coupled map with an even number of elements allows for equal partition into two clusters. This gives a plausible explanation for the smaller instability for even N system.)

system, the basin boundary of an attractor may move until, for a specific value of the parameter, the basin boundary touches the attractor. Then, if the attractor has a positive measure of initial conditions forming the basin of attraction, it becomes a Milnor attractor. Generally speaking, however, the above situation occurs only for very specific parameter values, and it is not naively expected that the Milnor attractors exist with a positive measure in the parameter space.

However, as was shown in the last section, Milnor attractors are found to be rather prevalent, occurring not only for specific isolated parameter values. Such dominance of Milnor attractors is often found in high-dimensional dynamical systems, for example coupled maps with 10 degrees of freedom or so. Then, the question we address now is why can there be so many Milnor attractors in a "high-dimensional" dynamical system, and what number of degrees of freedom is sufficient for constituting such 'high' dimensionality.

We computed the average basin fraction of Milnor attractors over the parameter interval $1.55 < a < 1.72$. In Fig.3, this fraction is plotted as a function of the number of degrees of freedom N. The increase of the average basin fraction of Milnor attractors with N is clearly visible for $N \approx (5 \sim 10)$, while it levels off for $N > 10$. Indeed, such increase of Milnor attractors with the degrees of freedom $5 \sim 10$ seems to be universal in a partially ordered phase in globally coupled chaotic system[16]. (Here, the degree of freedoms we use in the present paper is the number of units that has orbital instability. For example, if we choose a coupled system of N Lorenz equations, the degrees we mention is not $3N$, but N.)

Now we discuss a possible reason how the dominance of Milnor attractors appears. In a system with identical elements, due to the symmetry, there are at least $M(N_1, N_2, \cdots, N_k) = \frac{N!}{\prod_{i=1}^{k} N_i!} \prod_{\text{oversetsof} N_i = N_j} \frac{1}{m_\ell!}$ attractors for each clustering condition, where m_ℓ is the number of clusters with the same value N_j. Then, a combinatorial explosion in the number of attractors can be expected when many of the clustering conditions are allowed as attractors. For example, the permutation of N elements leads to $(N - 1)!$ possibilities and one might expect the number of attractors to be of this order. On the other hand, the phase space volume in a coupled system expands only exponentially with N. Typically the combinatorial explosion outruns the exponential increase around $N \approx (5 \sim 10)$. (For example compare 2^N and $(N - 1)!$. The latter surpasses the former at $N = 6$.) Hence, the attractors crowd[17] in the phase space and the stability of each attractor may be lost.

However, this argument seems to be incorrect. We have computed the number of attractors and compared with the basin fraction of Milnor attractors. As has been shown[16], the dominance of the Milnor attractors is not necessarily observed when the number of attractors is high. With increasing a, the fraction remains large even when the number of attractors has already decreased substantially. Since the basin volume of each of the attractors is far from being equal, the explosion in the number of attractors does not necessarily mean that the basin volume for each and every attractor should be very small. Indeed, according to our numerical results, for the parameter region where the Milnor attractors dominate, the number of Milnor attractors is not so high and the basin fraction of only a few Milnor attractors occupies almost all of phase space.

In the above sequence, the dominance of the Milnor attractors is observed when many attractors have disappeared. Therefore, we can revise the first explanation on the dominance by replacing the combinatorial explosion in the number of attractors themselves by the combinatorial explosion in the number of basin boundaries that separate the attractors. For the parameter region where many attractors start to disappear, there remain basin boundary points separating such (collapsed) attractors and the remaining attractors.

Now, we need to discuss how the distance between an attractor and its basin boundary changes with N. Consider a one-dimensional phase space, and a basin boundary that separates the regions of $x(1) > x^*$ and $x(1) < x^*$, while the attractor in concern exists at around $x(1) = x_A < x^*$, and the neighboring one at around $x(1) = x_B > x^*$. Now consider a region of N-dimensional phase space $x_A < x(i) < x_B$. If the region is partitioned by (basin) boundaries at $x(i) = x^*$ for $i = 1, \cdots N$, it is partitioned into 2^N units. Since this partition is just a direct product of the original partition by $x(1) = x^*$, the distance between each attractor and the basin boundary does not change with N. (For example, consider the extreme case that N identical maps are uncoupled ($\epsilon = 0$).)

On the other hand, consider a boundary given by some condition for $(x(1), \cdots, x(N))$, represented by a (possibly very complex) hyperplane $C(x(1), \cdots, x(N)) = 0$. In the present system with global (all-to-all) couplings, many of the permutational changes of $x(i)$ in the condition also give basin boundaries. Generally, the condition for the basin can also have clustering $(N_1, \cdots, N_k)$, since the attractors are clustered as such. Then the condition obtained by the permutation of $C(x(1), \cdots, x(N)) = 0$ gives a basin boundary also (or one can say that $C(x(1), \cdots, x(N)) = 0$ itself satisfies such permutational symmetry). Then, the basin boundary has $M(N_1, \cdots, N_k)$ segments transformed into each other by the permutations. The number of such segments of the boundaries increases combinatorially with N. Roughly speaking, the sum of $M(N_1, \cdots, N_k)$ increases in the order of $(N - 1)!$, when a variety of clusterings is allowed for the boundary. Now the N-dimensional phase space region is partitioned by $O(N - 1)!)$ basin boundary segments. Recalling that the distance between an attractor and the basin boundary remains at the same order for the partition of the order of 2^N, the distance should decrease if $(N - 1)!$ is larger than 2^N. Since for $N > 5$, the former increases drastically faster than the latter, the distance should decrease drastically for $N > 5$. Then for $N > 5$, the probability that a basin boundary touches with an attractor itself will be increased. Since this argument is applied for any attractors and their basin boundary characterized by complex clusterings having combinatorially large $M(N_1, \cdots, N_k)$, the probability that an attractor touches its basin boundary is drastically amplified for $N > 5$. Although this explanation may be rather rough, it gives a hint to why Milnor attractors are so dominant for $N \gtrsim (5 \sim 10)$. We surmise that this is the reason why Milnor attractors are dominant in our model at $(1.64 \sim 1.67)$ when $N \gtrsim (5 \sim 10)$.

Since the above discussion is based mainly on simple combinatorial arguments, it is expected that the dominance of Milnor attractors for $N \gtrsim (5 \sim 10)$ may be rather common at some parameter region in high dimensional dynamical systems. It is interesting that pulse-coupled oscillators with global coupling also show the prevalence of Milnor attractors for $N \geq 5$ [18].

One might expect that permutational symmetry us necessary for the prevalence of Milnor attractors. For example, in the GCM (1), the permutation symmetry arising from identical elements leads to a combinatorial explosion in the number of attractors. Then, one may wonder whether the prevalence of Milnor attractors is possible only for such highly symmetric systems. We have therefore studied a GCM with heterogeneous parameters. Although the fraction seems to be smaller than in the homogeneous case, Milnor attractors are again observed and their basin volume is rather large for some parameter region. As in the symmetric case, the basin fraction of Milnor attractors increases around $N \approx (5 \sim 10)$.

Note that even though complete synchronization between two elements is lost, clusterings as with regards to the phase relationships can exist. (As for such phase synchronization of chaotic elements see [19]). Indeed, there are two groups when considering the oscillations of phases as large-small-large... and small-large-small..., that are preserved in time for many attractors. Furthermore, finer preserved phase relationships can also exist. Similarly, it is natural to expect an explosion in the number of the basin boundary points for some parameter regime. Accordingly the argument on the dominance of Milnor attractors for a homogeneous GCM can be applied here to some degree as well.

It is also interesting to note that in Hamiltonian dynamical systems, agreement with thermodynamic behavior is often observed only for degrees of freedom higher than $5 \sim 10$[20]. Considering the combinatorial complexity woven by all the possible Arnold webs (that hence may be termed "Arnold spaghetti"), the entire phase space volume that expands only exponentially with the number of degrees of freedom may be covered by webs, resulting in uniformly chaotic behavior. If this argument holds, the degrees of freedom required for thermodynamic behavior can also be discussed in a similar manner.

4. Chaotic Itinerancy

In the PO phase, orbits often make itinerancy over several ordered states with partial synchronization of elements, through highly chaotic states. This dynamics, called chaotic itinerancy (CI), is a novel universal class in high-dimensional dynamical systems[21]. In the CI, an orbit successively itinerates over such "attractor-ruins", ordered motion with some coherence among elements. The motion at "attractor-ruins" is quasi-stationary. For example, if the effective degrees of freedom is two, the elements split into two groups, in each of which elements oscillate almost coherently. The system is in the vicinity of a two-clustered state, which, however, is not a stable attractor, but keeps attraction to its vicinity globally within the phase space. After staying at an attractor-ruin, an orbit eventually exits from it. This exit arises from orbital instability. In the above example, the synchronization among the two groups is increased. Then, as is straightforwardly seen in the model equation (1), the dynamics are approximately given by $x_{n+1} = f(x_n)$, which has stronger orbital instability than a clustered state.

With this instability the state enters into a high-dimensional chaotic motion
without clear coherence. (Here it is interesting to note that the effective
degrees of freedom decreases before it goes to a high-dimensional state).
This high-dimensional state is again quasi-stationary, although there are
some holes connecting to the attractor-ruins from it. Once the orbit is
trapped at a hole, it is suddenly attracted to one of attractor ruins, i.e.,
ordered states with low-dimensional dynamics.

This CI dynamics has independently been found in a model of neural
dynamics by Tsuda [22], optical turbulence [23], and in GCM[3, 24]. It
provides an example of successive changes of relationships among elements.

There seem to be several types of "chaotic itinerancy" covered by this
general definition for it. It can roughly be classified according to the degree
of correlation between the ordered states visited successively. The correlation is high if the paths for the transitions between the ordered states are
narrow, and the probabilities for visiting the next ordered state are rather
low. On the other hand, the correlation is low when the memory of the
previous sate is lost due to high-dimensional chaos during the transition.

Still, the systems with chaotic itinerancy studied so far commonly have
a small number of positive Lyapunov exponents and many exponents close
to zero. As a result, the dimension of the global attractor is high, while the
path in the phase space is restricted.

Note that the Milnor attractors satisfy the condition of the above ordered states constituting chaotic itinerancy. Some Milnor attractors we have
found keep global attraction, which is consistent with the observation that
the attraction to ordered states in chaotic itinerancy occurs globally from
a high-dimensional chaotic state. Attraction of an orbit to precisely a given
attractor requires infinite time, and before the orbit is really settled to a
given Milnor attractor, it may be kicked away. Then, the long-term dynamics can be constructed as the successive alternations to the attraction to,
and escapes from, Milnor attractors. If the attraction to robust attractors
from a given Milnor attractor is not possible, the long-term dynamics with
the noise strength $\to +0$ is represented by successive transitions over Milnor
attractors. Then the dynamics is represented by transition matrix over
among Milnor attractors. This matrix is generally asymmetric: often, there
is a connection from a Milnor attractor A to a Milnor attractor B, but
not from B to A. The total dynamics is represented by the motion over a
network, given by a set of directed graphs over Milnor attractors. In general,
the 'ordered states' in CI may not be exactly Milnor attractors but can be
weakly destabilized states from Milnor attractors. Still, the attribution of
CI to Milnor attractor network dynamics is expected to work as one ideal
limit.

Remark.   Computability of chaotic itinerancy has a serious problem, since switching process over Milnor attractor network in the noiseless case may differ from that of the case with the limit of noise $\rightarrow$ +0, or from that obtained by a digital computer with a finite precision. For example, once the digits of two variable $x(i) = x(j)$ agree down to the lowest bit, the values never split again, even though the state with the synchronization of the two elements may be unstable[25]. As long as digital computation is adopted, it is always possible that an orbit is trapped to such unstable state. (See [26], for one technique to resolve this numerical problem).

In each event of switching, which Milnor attractor is visited next after the departure from a Milnor attractor may depend on the precision, or on any small amount of noise. Here it may be interesting to note that there are similar statistical features between (Milnor attractor) dynamics with a riddled basin and undecidable dynamics of a universal Turing-machine[27].

5. Relevance to Neural Networks

When one considers (static) memory in terms of dynamical systems, it is often adopted that each memory is assigned into an attractor. Here, a system with many attractors is desirable as such system. Then, existence of Milnor attractors may lead us to suspect the correspondence between a (robust) attractor and memory.

Here, it may be interesting to recall that the term magic number 7 ± 2 was originally coined in psychology [28]. It was found that the number of chunks (items) that is memorized in short term memory is limited to 7 ± 2. Indeed with this number 7 ± 2, the fraction of basins for Milnor attractors increases. Since possible explanation is based only on combinatorial arguments, this 'magic number $5 \sim 10$' in dynamical systems does not strongly depend on the choice of specific models. Then, it may be interesting to discuss a possible connection of it with the original magic number 7 ± 2 in psychology. (see also [29] for a pioneering approach to this problem from a viewpoint of chaotic dynamics). To memorize k chunks of information including their order (e.g., a phone number of k digits) within a dynamical system, it is natural to assign each memorized state to an attractor of a k-dimensional dynamical system (unless rather elaborate mechanisms are assumed). In this k dimensional phase space, a combinatorial variety of attractors has to be presumed in order to assure a sufficient variety of memories. Then, if our argument so far is applied to the system, Milnor attractors may be dominant for $k > (5 \sim 10)$. If this is the case, the state represented by a Milnor attractor may be kicked out by tiny perturbations. Thus robust memory may not be possible for information that contains more than 7 ± 2 chunks. Possibly, this argument can also be applied to other systems that adopt attractors as memory, including most neural networks.

(Of course, the present argument should mainly be applied to systems with all-to-all couplings, or to highly connected network systems. If the connections are hierarchically ordered, the number of memory items can be increased. The often adopted module structure is relevant for this purpose.)

This argument does not necessarily imply that Milnor attractors are irrelevant to cognitive processes. For a dynamical system to work as a memory, some mechanism to write down and read it out is necessary. If the memory is given in a robust attractor, its information processing is not so easy, instead of its stability. Milnor attractors may provide dynamic memory [22, 1] allowing for interface between outside and inside, external inputs and internal representation. In a Milnor attractor, some structure is preserved, while it is dynamically connected with different attractors. Also, it can be switched to different memory by any small inputs. The connection to other attractors is neither one-to-one nor random. It is highly structured with some constraints.

Searches with chaos itinerating over attractor ruins has been discussed in[30, 22] with a support in an experiment on the olfactory bulb[30]. Freeman, through his experiments, proposed that the chaotic dynamics corresponds to a searching state for a variety of memories, represented by attractors [30], while evidence from human scalp EEG showing chaotic itinerancy is also suggested[31].

We note that the Milnor attractors in our GCM model provide a candidate for such a searching state, because of connection to a variety of stronger attractors which possibly play the role of rigidly memorized states. Stability of Milnor attractors by some noisy inputs also supports this correspondence.

Acknowledgements

The work is partially supported by Grant-in-Aids for Scientific Research from the Ministry of Education, Science, and Culture of Japan (11CE2006).

References

1. K. Kaneko and I. Tsuda *Complex Systems: Chaos and Beyond ──A Constructive Approach with Applications in Life Sciences* (Springer, 2000)
2. K. Kaneko, Physica 75 D (1994) 55
3. K. Kaneko, Physica 41 D (1990) 137-172
4. K. Kaneko, Prog. Theo. Phys. 72 (1984) 480-486; K.Kaneko ed., *Theory and applications of coupled map lattices, Wiley* (1993)
5. E. Mosekilde, Y. Maistrenko, and D. Postnov *Chaotic Synchronization*, World Scientific, 2002

6. K. Kaneko, J. Phys. A, 24 (1991) 2107

7. A. Crisanti, M. Falcioni, and A. Vulpiani, Phys. Rev. Lett. 76 (1996) 612; S.C Manruiba, A. Mikhailov, Europhys. Lett. 53(2001) 451-457

8. M. Mezard, G. Parisi, and M.A. Virasoro eds., *Spin Glass Theory and Beyond* (World Sci. Pub., Singapore, 1988)

9. K. Kaneko, Phys. Rev. Lett., 78 (1997) 2736-2739;

10. K. Kaneko, Physica D, 124 (1998) 322-344

11. J. Milnor, Comm. Math. Phys. 99 (1985) 177; 102 (1985) 517

12. P. Ashwin, J. Buescu, and I. Stuart, Phys. Lett. A 193 (1994) 126; Nonlinearity 9 (1996) 703

13. J.C. Sommerer and E. Ott., Nature 365 (1993) 138; E. Ott et al., Phys. Rev. Lett. 71 (1993) 4134

14. Y-C. Lai abd R.L.Winslow, Physica D 74 (1994) 353

15. See Fig.10 of ref. [10], where the right axis corresponding to the thick line should be read as $P(+0)$, not σ_c. Also the value 0.5 at the right axis of the right figure should be 1.0.

16. K. Kaneko, Phys. Rev. E.66 (2002) 055201(R)

17. The idea of this type of attractor crowding was first proposed by P. Hadley and K. Wiesenfeld (Phys. Rev. Lett. 62 (1989) 1335). However, it was later shown that states with different phase orderings (with $(N-1)!$ variety) are not separate attractors (K. Kaneko, Physica 55D (1992) 368; S. Watanabe and S. Strogatz, Phys. Rev. Lett. 70 (1993) 2391).

18. M. Timme, F. Wolf,and T. Geisel., Phys. Rev. Lett. 89 (2002) 154105

19. M.G. Rosenblum, A.S. Pikovsky, and K. Kurths, Phys. Rev. Lett 76 (1996) 1804 (see also K. Kaneko, Physica 37D (1989)60). Clusterings only as to the phases of oscillations are its natural extension.

20. S. Sasa and T.S. Komatsu, Phys. Rev. Lett. 82 (1999) 912: N. Nakagawa and K. Kaneko, Phys. Rev. E 64(2001) 055205(R)-209:

21. K. Kaneko and I. Tsuda, ed., Focus issue of Chaotic itinerancy, Chaos (2003), to appear,

22. I. Tsuda, World Futures 32(1991)167; Neural Networks 5(1992)313

23. K. Ikeda, K. Matsumoto, and K. Ohtsuka, Prog. Theor. Phys. Suppl. 99 (1989) 295

24. K. Kaneko, Physica 54 D (1991) 5-19

25. K. Kaneko, Physica D 77 (1994) 456

26. A. Pikovsky , O. Popovych, and Y. Maistrenko Phys. Rev. Lett. 87 (2001) 4102

27. A. Saito and K. Kaneko, Physica D, 155 (2001) 1-33

28. G.A. Miller, *The psychology of communication*, 1975, Basic Books, N.Y.

29. J.S.Nicolis and I.Tsuda, Bull. Math. Biol. 47(1985)343.

30. W. Freeman and C. A. Skarda, Brain Res. Rev. 10 (1985) 147; Physica D 75 (1994)151.

31. W. Freeman, in [21]

GENERALIZATION OF THE FEIGENBAUM-KADANOFF-SHENKER RENORMALIZATION AND CRITICAL PHENOMENA ASSOCIATED WITH THE GOLDEN MEAN QUASIPERIODICITY

S. P. KUZNETSOV
Saratov Division of Institute of Radio-Engineering and Electronics, Russian Academy of Sciences, Zelenaya 38, Saratov, 410019, Russia

Abstract

The paper presents a two-dimensional version of the Feigenbaum-Kadanoff-Shenker renormalization group equation. Several universality classes of critical behavior are discussed, which may occur at the onset of chaotic or strange nonchaotic attractors via quasiperiodicity at the golden-mean frequency ratio. Parameter space arrangement and respective scaling properties are discussed and illustrated.

1. Introduction

In modern nonlinear dynamics the concept of synchronization is considered not only in a classic sense, as a periodicity in a motion of a self-oscillator induced by the driven force, but relates to a variety of situations when dynamics of autonomous systems or external force are chaotic, quasiperiodic etc. It is used to speak of synchronization in a generalized sense if the system reproduces some definite features of the time-dependence of the external force in its dynamics. (See [1] and references therein.)

In a context of multi-parameter analysis, domains of synchronization, as well as the bifurcation sets, may be thought geometrically, as some configurations in parameter space. To understand the parameter space structure, it is essential to reveal critical situations responsible for formation or destruction of synchronized and non-synchronized regimes, in particular, those associated with birth of chaos or strange nonchaotic attractors (SNA)

79

A. Pikovsky and Y. Maistrenko (eds.), Synchronization: Theory and Application, 79–100.

[2,3,4]. Apparently, as a rule, the critical situations allow analysis in terms of renormalization group (RG) approach, analogous to that developed by Feigenbaum for the period-doubling transition to chaos [5,6], and they are characterized by properties of universality and scaling specific to each type of criticality. The critical situations are classified naturally in order of their codimension, which is a minimal number of parameters to be adjusted to reach a critical situation under study.

One of the models traditionally used for analysis of synchronization and desynchronization is the classic circle map

$$x_{n+1} = x_n + r + (K/2\pi)\sin 2\pi x_n. \tag{1}$$

In the parameter plane of this map (r, K) a set of synchronization regions – Arnold tongues is present, and they approach the axis $K=0$ by their sharp edges at rational points of r. On the line $K=1$ a critical point exists that corresponds to destruction of quasiperiodic regime with the golden-mean rotation number $w = \lim x_n = (\sqrt{5} - 1)/2$. Fine structure of the synchronization tongues and quasiperiodic regions near this point obeys universality and scaling properties deduced from the RG analysis of Feigenbaum, Kadanoff, Shenker [7], and Rand, Ostlund, Satija, Siggia [8,9]. We will refer to it as the *GM critical point* (GM stands for the 'golden-mean').

In the present paper we consider and discuss a generalized RG approach, which includes the critical behavior of GM type as a particular case. It opens a possibility to reveal and study new universality and scaling classes linked with a birth of SNA in quasiperiodically forced systems at the golden-mean frequency ratio. The approach is based on the two-dimensional version of the Feigenbaum-Kadanoff-Shenker equation.

In Sec. 2 the procedure of RG analysis appropriate for the golden-mean quasiperiodicity is explained, and a two-dimensional generalization of the approach of Feigenbaum-Kadanoff-Shenker [7] and Ostlund et al. [8,9] is developed. In Sec. 3 we discuss model systems including quasiperiodically driven logistic, circle, and fractional-linear maps. In Sec. 4 our generalized RG scheme is used to reproduce some results of classic analysis of quasiperiodic transition to chaos in the circle map. In Sections 5, 6, and 7 we review three novel types of critical behavior discovered in a course of joint research program with the group of nonlinear dynamics and statistical physics from Potsdam University (A. Pikovsky, U. Feudel, E. Neumann) [10,11,12]. For each type of criticality we illustrate scaling for the critical attractor associated with dynamics exactly at the critical point, and scaling of topography of the parameter plane near the criticality.

2. Two-dimensional generalization of the Feigenbaim-Kadanoff-Shenker equation

Let us consider quasiperiodic dynamics in some system with two basic frequencies, ω_1 and ω_2, and assume that two subsystems associated with these frequencies are coupled unidirectionally. To describe dynamics in terms of Poincaré map, we perform stroboscopic cross-section of the extended phase space by planes of constant time, separated by $T = 2\pi/\omega_2$. The first subsystem ("master") is independent of the second one, and the associated dynamical variable is the phase φ governed by equation $\varphi_{n+1} = \varphi_n + \omega_1 T$ (mod 2π). For the second subsystem ("slave") we assume that the dynamics is essentially one-dimensional: $x_{n+1} = F(x_n, \varphi_n)$. In respect to the second argument the function $F(x, \varphi)$ is 2π-periodic. Instead of φ we introduce a variable u defined modulo 1:

$$x_{n+1} = f(x_n, u_n), \ u_{n+1} = u_n + w \ (\text{mod} 1), \tag{2}$$

where $f(x, u) = F(x, 2\pi u)$, $w = \omega_1 T/2\pi = \omega_1/\omega_2$. In the further study we fix $w = (\sqrt{5} - 1)/2$.

In general context of nonlinear dynamics, the basic idea of the RG analysis consists in the following. We start with an evolution operator of a system on a definite time interval and apply this operator several times to construct the evolution operator for larger interval. Then, we try to adjust parameters of the original system to make the new operator reducible to the old one by scale change of dynamical variables. This procedure is called the RG transformation. The adjusted parameters will define location of the critical point. The RG transformation may be applied again and again to obtain a sequence of the evolution operators for larger and larger time intervals.

If the approach works, one possibility is that the produced operators become asymptotically identical, and we speak about a fixed point of the RG transformation. Another possibility is that they repeat each other after several steps of the RG transformation, and we speak about a periodic orbit, or a cycle of the RG equation. In any of these cases, the rescaled long-time evolution operators will be determined by structure of the RG transformation, rather than by concrete dynamical equations of the original dynamical system. This implies *universality*. On the other hand, repetition of the rescaled evolution operators at subsequent steps of the RG transformation means that the system manifests similar dynamics on different time scales. This implies *scaling*.

How can we apply this approach to critical phenomena associated with the golden-mean quasiperiodicity? As known, the convergent sequence of rationals for $w = (\sqrt{5} - 1)/2$ is defined as F_{k-1}/F_k, F_k are the Fibonacci numbers: $F_0 = 0, F_1 = 1, F_{k+1} = F_k + F_k$. This sequence delivers the best possible approximation for w, so, the dynamics on a time interval F_k is close to periodic. Hence, it is natural to consider a sequence of evolution operators over intervals of discrete time given by the Fibonacci numbers.

Let $f^{F_k}(x, u)$ and $f^{F_{k+1}}(x, u)$ designate transformation of x after F_k and F_{k+1} iterations, respectively. To construct the next operator, for F_{k+2} iterations, we start from (x, u) and perform first F_{k+1} iterations to arrive at $(f^{F_{k+1}}(x, u),\ u + F_{k+1}w)$, and then the rest F_k iterations with the result

$$f^{F_{k+2}}(x, u) = f^{F_k}(f^{F_{k+1}}(x, u),\ u + wF_{k+1}). \qquad (3)$$

To have a reasonable limit behavior of the evolution operators we change scales for x and u by some factors α and β at each new step of the construction, and define the renormalized functions as

$$g_k(x, u) = \alpha^k f(x/\alpha^k, u/\beta^k). \qquad (4)$$

Note that $wF_{k+1} = -(-w)^{k+1}$ (mod 1), so it is natural to set $\beta = -1/w = -1.618034...$ Rewriting (3) in terms of the renormalized functions we come to the functional equation

$$g_{k+2}(x, u) = \alpha^2 g_k(\alpha^{-1} g_{k+1}(x/\alpha, -uw), w^2 u + w). \qquad (5)$$

In the present article we deal with several different solutions of this equation – fixed points or cycles in the functional space. The constant α is specific for each universality class; it is evaluated in a course of solution of the functional equation.

The next step of the RG analysis consists in the following. Let us suppose that we deal now with dynamics in a vicinity of the critical point it in the parameter space. Then, a perturbation of the solution appears. Analyzing evolution of the perturbation of the evolution operators under subsequent application of the RG transformation we come to an eigenvalue problem. A number of relevant eigenvalues define a *codimension* of the critical situation. The relevant eigenvalues are those, which are larger than 1 in modulus, are not associated with infinitesimal variable changes, and do not violate the commutative properties of successively applied evolution operators (see e.g. [7-12] for some details). The codimension may be understood as a number of parameters, which must be adjusted to reach the criticality. For instance, in three-dimensional parameter space the codimension-one situations may occur at some surfaces, codimension-two situations at curves, and codimension-three at some points.

To derive an explicit form of the linearized RG equation appropriate for a vicinity of a fixed point $g(x, u)$ we substitute $g_k(x, u) = g(x, u) + \varepsilon h_k(x, u)$, $\varepsilon \ll 1$ and account terms of the first order in ε in Eq.(5). Then, setting

$h_k(x) = \delta^k h(x)$ we arrive to the eigenvalue problem

$$\delta^2 h(x, u) = \alpha \delta g'(g(x/\alpha, -uw), w^2 u + w) h(x/\alpha, -uw) + \alpha^2 h(\alpha^{-1} g(x/\alpha, -uw), w^2 u + w). \tag{6}$$

For each particular type of criticality, with specific $g(x, u)$ and α, this equation can be solved numerically to obtain spectrum of relevant δ.

3. Basic models

The simplest example, for which the developed RG scheme is applicable, is the well-known circle map (1). Figure 1 shows a chart of dynamical regimes

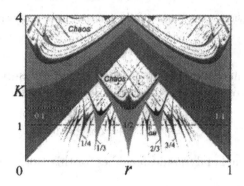

Figure 1. Chart of dynamical regimes on the parameter plane of the circle map. Numbers inside Arnold tongues indicate the respective rotation numbers.

on the parameter plane (r, K). For $K < 1$ one can observe periodic or quasiperiodic regimes associated with rational or irrational values of the rotation number defined as $\rho(r, K) = \lim_{n \to \infty} x_n/n$. Periodic regimes take place inside the Arnold tongues, and quasiperiodic motions are observed between of them. One can find a curve of constant golden-mean rotation number: $\rho(r, K) = w$. This curve starts at $K = 0$, $r = w$, and meets the critical line $K = 1$ at the GM critical point

$$K_{GM} = 1, r_{GM} = 0.60666106347\ldots \tag{7}$$

discovered by Shenker [13] and afterwards studied in terms of RG analysis by Feigenbaum–Kadanoff–Shenker and by Ostlund et al. [7-9].

Further examples of types of critical behavior discussed in the present article occur in quasiperiodically forced maps.

One model is the quasiperiodically driven logistic map [14-18, 10, 11]. A usual logistic map $x_{n+1} = \lambda - x_n^2$ is a basic model to study period-doubling transition to chaos. As it has the only relevant parameter λ, a natural generalization for presence of the external driving is to assume that this

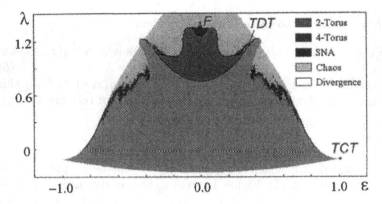

Figure 2. Chart of dynamical regimes on the parameter plane of quasiperiodically driven logistic map (8).

parameter is modulated with some frequency. In our study this frequency, measured in units of time discretization, is fixed, $w = (\sqrt{5} - 1)/2$. So, the model is

$$x_{n+1} = \lambda - x_n^2 + \varepsilon \cos 2\pi n w. \tag{8}$$

Figure 2 shows a chart of dynamical regimes for this model on the parameter plane $(\varepsilon, \quad \lambda)$.

For $\varepsilon = 0$ Eq.(8) becomes the conventional logistic map. So, what is observed along the line $\varepsilon = 0$ is the usual period-doubling cascade, accumulated to the limit critical point of Feigenbaum (point F) [5,6,31].

First, let us take a value of λ at which the unforced map has a stable fixed point. At nonzero ε the fixed point will be transformed into a stable smooth invariant curve. In continuous-time dynamical systems such curves appear in the Poincaré cross-section for the motion on tori, so, with commonly used abuse of the terminology, we speak about the torus-attractor T1.

If the external force of small amplitude drives a stable period-2 orbit, it gives rise to an attractor consisting of two closed smooth curves, the doubled torus T2. Period-4 orbit generates, respectively, a four-piece invariant curve (torus T4), and so forth. In contrast to usual period-doubling, the sequence of torus-doubling transitions appears to be finite: the smaller amplitude of driving, the larger number of torus doublings seen in a course of increase of λ [14-20].

If we keep λ constant and increase the force amplitude, the smooth torus may transform into SNA: the Lyapunov exponent remains negative, but the geometrical structure of the attractor becomes complex, fractal-like. Also regimes with positive Lyapunov exponent arise for larger λ and

ε. With further increase of the parameters the orbits escape to infinity (white domain in Fig. 2).

As known, the parameter interval corresponding to existence of an attractive fixed point in the unforced logistic map $\lambda \in (-0.25, 0.75)$ is bounded on one side by the tangent bifurcation, collision of a pair of fixed points (stable and unstable), with their subsequent disappearance. On the other side it is bounded by the period-doubling bifurcation. Analogously, the bottom border of the domain T1 in Fig. 2 is the bifurcation curve of tori collision: attractor and repeller, represented by two invariant curves, approach each other, collide, and disappear. The top border is the bifurcation curve of torus doubling: the attractor originates here consisting of two closely placed curves, and after the bifurcation they move one off another.

Let us start at $\varepsilon = 0$, $\lambda = -0.25$ and increase ε to go in the parameter plane along the torus collision bifurcation curve. The situation of collision of smooth invariant curves takes place while the motion is confined on one side of the logistic parabola. At some value of ε the invariant curve at the bifurcation threshold touches the extremum $x = 0$. In accordance with argumentation of Ref.[11], it corresponds to the terminal point of the bifurcation line. This is a critical situation of particular interest, the TCT critical point (TCT stands for 'torus collision terminal') [11]:

$$\lambda_{TCT} = -0.09977122895\ldots, \varepsilon_{TCT} = 1.01105609099\ldots. \qquad (9)$$

Now, let us start at $\varepsilon = 0$, $\lambda = 0.75$ and move along the torus-doubling bifurcation curve. As in the previous case, this bifurcation of smooth invariant curve takes place only while the whole curve is placed on one side of the logistic parabola. At some value of ε the invariant curve at the bifurcation threshold touches the extremum $x = 0$, and the torus-doubling bifurcation line is terminated. This is the TDT critical point (TDT stands for 'torus-doubling terminal') [10]:

$$\lambda_{TDT} = 1.158096856726\ldots, \quad \varepsilon_{TDT} = 0.360248020507\ldots. \qquad (10)$$

TCT and TDT critical points were found also in quasiperiodically forced circle map

$$x_{n+1} = x_n + r - (K/2\pi)\sin 2\pi x_n + \varepsilon \cos 2\pi n w \; (\mathrm{mod}\,1) \qquad (11)$$

in the supercritical case $K > 1$ (near the extrema it looks locally like the logistic map). In some respects, this is a more convenient object for detailed study: no divergence can occur in this map because the variable x is defined modulo 1.

Figure 3 shows a chart of dynamical regimes for the driven circle map on a part of the parameter plane (b, ε) including the TCT critical point [11].

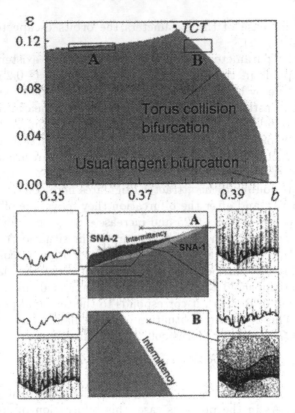

Figure 3. Chart of dynamical regimes on the parameter plane (b, ε) and two enlarged fragments with phase portraits of attractors on phase plane (u, x) at representative points.

Separately, two rectangular fragments of the chart are shown, together with phase portraits of attractors at representative points.

The large gray domain in the diagram corresponds to existence of the localized torus attractor. The right border of this domain is the bifurcation curve of bifurcation of collision of a pair of smooth tori, one stable and another unstable. After the event, both of them disappear, and intermittent regime occurs, with long-time travel of the orbits through the region of former existence of the tori (the 'channel'). Going along the bifurcation curve we observe that the semi-attractive invariant curve, formed at the moment of the collision, grows in size, and ultimately touches the minimum of the map; there we arrive at the TCT point. As found numerically, it is located at

$$r_{TCT} = 0.377866239..., \quad \varepsilon_{TCT} = 0.132566321... \tag{12}$$

Another, upper border of the gray area corresponds to a situation when the stable and unstable invariant curves touch each other, but do not co-incide. This means that at least one of the curves must be non-smooth ('fractal torus'). From the figure one can see that both bifurcation lines of smooth and fractal tori-collision meet at the TCT critical point.

It was observed that fractalization of torus and transition to SNA in the forced circle map is possible also in the critical and subcrital domain ($K \leq 1$) [21,22]. This transition can not be associated with the TDT or TCT points because of absence of a quadratic exteremum. The nature of the criticality was revealed in Ref. [12] as linked with the torus fractalization at the intermittency threshold. To describe the phenomenon a model was used

$$x_{n+1} = f(x_n) + b + \varepsilon \cos 2\pi w n, \tag{13}$$

with $f(x)$ defined as

$$f(x) = \begin{cases} x/(1-x), & x \leq 0.75 \\ 9/2x - 3, & x > 0.75 \end{cases} \tag{14}$$

One branch of the mapping is selected in a form of the fractional-linear function, $x/(1-x)$, which appears naturally in analysis of dynamics near the tangent bifurcation associated with intermittency (e.g. [23-26]). The other branch is attached somewhat arbitrarily to ensure presence of the 're-injection mechanism' in the dynamics and to exclude divergence.

Figure 4 shows a chart of dynamical regimes for the model (13). The white area designates chaotic regime with positive Lyapunov exponent Λ. Gray regions correspond to negative Λ. In the bottom gray area attractor is localized and represented by a smooth torus. The upper border of this region is the bifurcation curve of transition to a delocalized attractor via intermittency. The bifurcation consists in collision of smooth stable and unstable tori with their coincidence, and the Lyapunov exponent at the bifurcation is zero. In the right-hand part of the diagram the bifurcation curve separates regimes of torus and SNA. The bifurcation corresponds to a fractal collision of two invariant curves at some exceptional set of points, and the Lyapunov exponent at the bifurcation is negative. These two parts of the bifurcation border are separated by *the critical point of torus fractalization* (TF) located at

$$\varepsilon_{TF} = 2, \; b_{TF} = -0.597515185376121\ldots \tag{15}$$

4. The classic GM critical point

Critical behavior in the circle map associated with break-up of the golden-mean quasiperiodicity (GM critical point) was discovered first by Shenker

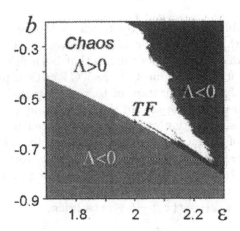

Figure 4.   Chart of dynamical regimes for the model (13). The bottom gray area corresponds to localized attractor represented by smooth torus. The upper border is the bifurcation curve of the intermittent transition. In the left part the bifurcation consists in collision of smooth stable and unstable tori with their coincidence, in the right part – to fractal collision at some exceptional set of points. White area designates chaos, and dark gray presumably corresponds to SNA. Sign of the Lyapunov exponent Λ is indicated in all three domains.

[13] and studied in terms of RG analysis by Feigenbaum–Kadanoff–Shenker and Ostlund et al. [7-9]. Although the circle map is one-dimensional, it may be treated in terms of our general scheme, as a particular case of (1). We consider two decoupled maps

$$x_{n+1} = f(x_n), \quad u_{n+1} = u_n + w \pmod 1, \tag{16}$$

with $f(x) = x + r - (K/2\pi)\sin 2\pi x$. The function is independent of the second argument u, so, the GM criticality will correspond to a degenerate fixed point of our functional equation: $g_k(x, u) \equiv G(x)$. In this case Eq.(5) yields

$$G(x) = \alpha^2 G(\alpha^{-1} G(x/\alpha)), \tag{17}$$

the relation known as the Feigenbaum–Kadanoff–Shenker equation. It has been solved numerically (e.g. [7-9, 27-30]), and the function is found in a form of high-precision expansion in powers of x^3. The scaling constant is

$$\alpha = -1.288574553954\ldots \tag{18}$$

Accounting representation of the circle map in the form (16) it is natural to depict the critical attractor in coordinates (u, x) (Fig. 5). Observe that it is represented by a fractal-like curve. Locally, the basic scaling property of this fractal may be deduced from the RG analysis. Indeed, the evolution operators for time intervals increasing as Fibonacci numbers, become

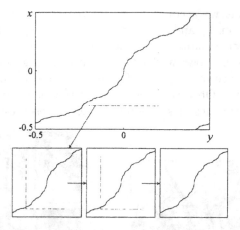

Figure 5.   Attractor of the two-dimensional map (16) at the GM critical point (top panel) and illustration of the basic local scaling property: the structure reproduces itself under magnification with factors $a = -1.28857$ and $b = -1.61803$ along the vertical and the horizontal axes, respectively.

identical, up to the scale change. For each next Fibonacci number the variables x and u are rescaled by α and $\beta = -w^{-1}$. As follows, the attractor in coordinates (u, x) must possess self-similarity: increasing resolution by factors α and β along the vertical and the horizontal axes, respectively, one should observe the similar structures (see bottom panels of Fig. 5).

For perturbations of the GM fixed-point, which do not violate the unidirectional nature of the master-slave coupling, the equation (6) accepts the form

$$\delta^2 h(x) = \alpha \delta G'(G(x/\alpha))h(x/\alpha) + \alpha^2 h(\alpha^{-1}G(x/\alpha)). \qquad (19)$$

As found (e.g. Refs. [7-9,27-30]), there are two relevant eigenvalues,

$$\delta_1 = -2.8336106559... \text{ and } \delta_2 = \alpha^2 = 1.660424381... \qquad (20)$$

These constants are responsible for the scaling properties of the parameter space structure near the GM critical point. However, to demonstrate the scaling we need to define a special local coordinate system near the critical point – the scaling coordinates. (The same will be necessary for other types of criticality, see sections 5-7.) As argued in Refs.[29,30], this is a curvilinear system: one coordinate line goes along the critical line $k = 1$, and another along the curve of constant rotation number. Numerically, a link of new coordinates (C_1, C_2) with the parameters of the original map is expressed as

$$r = r_c + c_1 - 0.01749c_2 - 0.00148c_2^2, \quad k = k_c + c_2. \qquad (21)$$

In these equations we account terms up to the second order because of the relation between δ_1 and δ_2: $\delta_2 < \delta_1$ and $\delta_2 < \delta_1^2$, but $\delta_2 > \delta_1^3$ (see Refs. [31,32,11,12] for explanation of the rules for selection of the scaling coordinates). Figure 6 shows a chart of dynamical regimes with Arnold tongues and a sequence of fragments for several steps of magnification in the scaling coordinates. Observe excellent repetition of the two-dimensional arrangement of the tongues at subsequent levels of resolution.

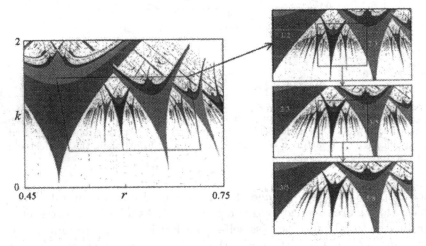

Figure 6.   Chart of dynamical regimes on the parameter plane of the sine circle map and a sequence of fragments for several steps of magnification of vicinity of the GM critical point in the scaling coordinates, with factors δ_1 and δ_2 along horizontal and vertical axes, respectively.

5. Critical point TCT

RG analysis of the torus-collision terminal point was developed in Ref. [12]. The critical behavior of this type was found in the forced logistic map (8) and in the forced supercritical circle map (11). Here we prefer to deal with the last one because divergence of iterations is excluded for sure in this case. The equation may be written as

$$x_{n+1} = x_n + r - (K/2\pi)\sin 2\pi x_n + \varepsilon \cos(2\pi u) \ (\mathrm{mod}\,1),$$
$$u_{n+1} = u_n + w \ (\mathrm{mod}\,1), \tag{22}$$

and parameter K is supposed to be supercritical and fixed, $K = 2.5$. As mentioned in Sec.2, the TCT point is located at

$$(r, \ \varepsilon)_{TCT} = (0.377866239, \ 0.132566321).$$

In the RG approach, the TCT point is associated with a fixed-point solution of the functional equation (5). This circumstance was checked accurately in the numerical procedure based on iterations of the RG transformation (5). Also the multi-dimensional Newton technique was used to solve the fixed-point equation in respect to the coefficients of polynomial expansion of the universal function in an appropriately chosen domain in the (u, x) plane (see [12] for details). The scaling constant α was found in the course of the computations, so

$$\alpha = 1.7109605\ldots \text{ and } \beta == 1.6180339\ldots \tag{23}$$

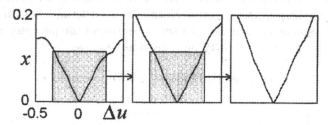

Figure 7. Attractor of the forced circle map at the TCT critical point (the left panel) and illustration of the basic local scaling property: the structure reproduces itself under magnification with factors $\alpha = 1.71096$ and $\beta = -1.61803$ along the vertical and the horizontal axes, respectively.

As seen from Fig. 7, the critical attractor in coordinates (u, x) is represented by a non-smooth fractal-like curve. To observe scaling, we need to select properly the origin of local coordinate system (the 'scaling center'). As found in Ref. [12], it has to be placed at

$$u_c = 0.284109286 \text{ and } x_c = (2\pi)^{-1} \arctan \sqrt{K^2 - 1} = 0.184505060. \tag{24}$$

Now, if we rescale $\Delta x = x - x_c$ and $\Delta u = u - u_c$ by factors α and $\beta = -w^{-1}$, respectively, the dynamical regimes remain of the same kind, but with rescaling of time by factor w^{-1}. The attractor also must be invariant under this transformation. Indeed, the picture inside a selected box in Fig. 7 reproduces itself under subsequent magnifications (with inversion in respect to the phase variable, due to the negative β). This scaling property implies that locally the behavior of the invariant curve obeys $\Delta x \propto |\Delta u|^\gamma$ with $\gamma = \log \alpha / \log \beta \cong 1.117$. The power γ is close to one, so visually the curve looks like broken at the point of singularity. Due to ergodicity ensured by irrationality of the frequency ratio, the singularity at the origin implies existence of the same type of singularities over the whole invariant curve, in a dense set of points. Note that $\gamma > 1$. It means that the singularity is weak: the invariant curve, apparently, remains differentiable, but not twice differentiable.

The next step is analysis of the linearized RG equation and of the corresponding eigenvalue problem (6). Numerical solution of the functional equation with substitution of $g(x, u)$ and constant α associated with the TCT criticality was performed, with approximation of the eigenfunctions via finite power expansions in respect to x and u. As found, two eigenvalues are relevant:

$$\delta_1 = 3.600810\ldots \text{ and } \delta_2 - 1.828329\ldots \qquad (25)$$

These are scaling factors determining self-similarity of topography in a vicinity of the TCT point. To demonstrate the scaling property we define scaling coordinates in the parameter plane. Note that $\delta_2 < \delta_1$ and $\delta_2 < \delta_1^2$, but $\delta_2 > \delta_1^3$. So, we account terms up to the second order in the parameter change. As suggested in Ref.[12], it may be chosen as

$$r = r_c + c_1 - 0.3121848c_2 - 2.047c_2^2, \quad \varepsilon = \varepsilon_c + c_2. \qquad (26)$$

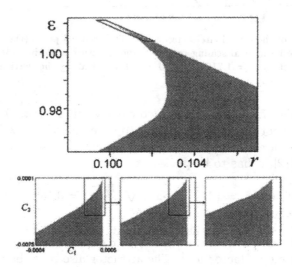

Figure 8. Chart of dynamical regimes on the parameter plane of the quasiperiodically driven supercritical circle map and a sequence of fragments for several steps of magnification of a vicinity of the TCT critical point in the scaling coordinates, with factors δ_1 and δ_2 along horizontal and vertical axes, respectively. Gray area corresponds to localized attractor with negative Lyapunov exponent, and white to chaos.

Figure 8 shows a fragment of the chart of dynamical regimes near the TCT point for the forced circle map. Note similarity of the configurations represented in the scaling coordinates.

6. Critical point TDT

Let us turn now to the RG results for the torus-doubling terminal point
[10,33]. The basic illustrative example will be the forced logistic map that
may be rewritten as

$$x_{n+1} = \lambda - x_n^2 + \varepsilon \cos 2\pi u_n, \quad u_{n+1} = u_n + w \ (\mathrm{mod}\,1). \tag{27}$$

As noted in Sec. 2, the TDT point is located at
$(\lambda, \ \varepsilon)_{TDT} = (1.158096856, 0.360248020)$.

It was found in Refs.[10,33] that the TDT point is associated with a
period-3 cycle of the RG equation (5): $g_1(x,u) \to g_2(x,u) \to g_3(x,u) \to$
$g_1(x,u)$. To find this period-3 solution with high precision a numerical
procedure was developed. The result is a representation of functional pair
$\{g_1(x,u), g_2(x,u)\}$ in a form of polynomial expansions over the arguments
x and u (see the table of coefficients in [33]). The rescaling constant is
$\alpha = 1.58259341 \ldots$

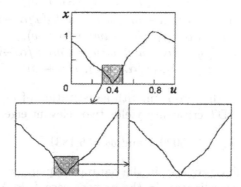

Figure 9.   Attractor of the forced logistic map at the TDT critical point and illustration
of the basic local scaling property: the structure reproduces itself under magnification
with factors $\alpha = 3.96376$ and $\beta = -4.2360$ along the vertical and the horizontal axes,
respectively.

In coordinates (u, x) the critical attractor looks like a fractal curve
(Fig. 9). To observe scaling, the origin of the coordinate system must be
placed at the 'scaling center' [10,33]

$$u_c = 0.3952188264 \text{ and } x_c = 0. \tag{28}$$

Due to the period-3 nature of the solution of the RG equation, observation of self-similarity of the critical attractor requires using the scaling
factors

$$\alpha^3 = 3.96376647 \ldots \text{ and } \beta^3 = -4.23606798 \ldots \tag{29}$$

If we rescale x and $\Delta u = u - u_c$ by α^3 and β^3, respectively, the dynamical regimes remain of the same kind, but with characteristic time rescaled by w^{-3}. The curve representing the attractor must be invariant under this transformation, and this is indeed the case, see Fig. 9. The picture inside a selected box reproduces itself under subsequent magnifications. Locally the invariant curve behaves as $x \propto |\Delta u|^\gamma$ with $\gamma = \log \alpha / \log \beta \cong 0.954$. The exponent is close to one, so the curve looks like broken at the point of singularity. Due to ergodicity of the quasiperiodic motion, the singularity at the origin implies presence of the same type of singularities on a dense set of points over the whole invariant curve. As $\gamma < 1$, the curve is not differentiable.

Because of the period-3 nature of the RG equation solution, analysis of the linearized RG equation is more complicated than that for a fixed point. The eigenvalue problem reads

$$
\begin{aligned}
\delta^2 h_3(x,u) &= \alpha \delta g_1'(g_2(x/\alpha, -uw), w^2 u + w) h_2(x/\alpha, -uw) \\
&\quad + \alpha^2 h_1(\alpha^{-1} g_2(x/\alpha, -uw), w^2 u + w), \\
\delta^2 h_1(x,u) &= \alpha \delta g_2'(g_3(x/\alpha, -uw), w^2 u + w) h_3(x/\alpha, -uw) \\
&\quad + \alpha^2 h_2(\alpha^{-1} g_3(x/\alpha, -uw), w^2 u + w), \\
\delta^2 h_2(x,u) &= \alpha \delta g_3'(g_1(x/\alpha, -uw), w^2 u + w) h_1(x/\alpha, -uw) \\
&\quad + \alpha^2 h_3(\alpha^{-1} g_1(x/\alpha, -uw), w^2 u + w).
\end{aligned}
\tag{30}
$$

Numerical solution of this problem with substitution of $g_{1,2,3}(x,u)$ and α associated with the TDT criticality yields two relevant eigenvalues [10,33]:

$$
\delta_1 = 10.5029... \quad \text{and} \quad \delta_2 = 5.1881... \tag{31}
$$

To demonstrate scaling property in the parameter plane we need to define appropriate scaling coordinates. In the present case $\delta_2 < \delta_1$ and $\delta_2 > \delta_1^m$ for $m = 2, 3, \ldots$. It means that a linear parameter change is sufficient. According to Refs.[10,33], it may be chosen as

$$
\lambda = \lambda_{TDT} + c_2, \quad \varepsilon = \varepsilon_{TDT} - c_1 + 0.3347 c_2. \tag{32}
$$

Figure 10 shows a chart of dynamical regimes near the TDT in scaling coordinates for several steps of subsequent magnification.

7. Critical point TF

The transition from localized to delocalized attractor in the model map (13) is accompanied by appearance of intermittent regimes. While we are close to the point of bifurcation, the laminar stages of dynamics occupy an overwhelming part of observation time (like in the case of the usual

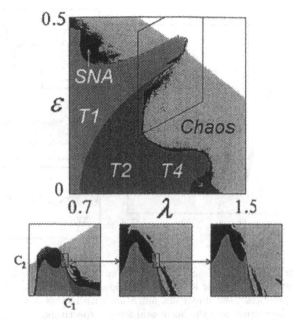

Figure 10.   Chart of dynamical regimes on the parameter plane of the quasiperiodically driven logistic map and a sequence of fragments for several steps of magnification of a vicinity of the TDT critical point in the scaling coordinates, with factors δ_1 and δ_2 along horizontal and vertical axes, respectively. Gray area corresponds to localized attractor with negative Lyapunov exponent, and white to chaos.

Pomeau-Manneville intermittency). They correspond to dynamics on the left branch of the map (13). To study details of the transition we may concentrate on the laminar stages and consider a simplified map

$$x_{n+1} = x_n/(1-x_n) + b + \varepsilon \cos(2\pi(nw+u)), \quad u_{n+1} = u_n + w \,(\bmod\,1). \quad (33)$$

As explained in Sec.2, the bifurcation border in the plane (ε, b) contains a critical point TF separating situations of smooth and fractal tori collision at $(\varepsilon, b)_{TF} = (2, -0.597515185)$.

An important note is that due to the fractional-linear nature of the map the functions obtained at subsequent steps of the RG transformation (5) will be fractional-linear too. The same is true for the fixed-point of the RG equation, associated with the TF critical point. It implies that we may search a solution for the fixed-point in a form

$$g(x, u) = (a(u)x + b(u))/(c(u)x + d(u)), \quad (34)$$

where a, b, c, d are some functions of u. Without loss of generality we require them to satisfy additional conditions $a(u)d(u) - b(u)c(u) \equiv 1$ and $c(0) = 1$.

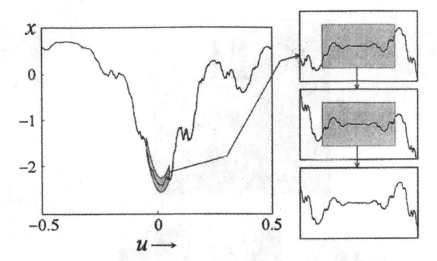

Figure 11.   Attractor of the forced fractional-linear map at the TF critical point (the left panel) and illustration of the basic local scaling property: the structure depicted in scaling coordinates reproduces itself under magnification with factors $\alpha = 2.89005$ and $\beta = -1.618034$ along the vertical and the horizontal axes, respectively.

Substituting (34) into (5) we arrive at the fixed-point RG equation in terms of the functions a, b, c, d:

$$\begin{pmatrix} a(u) & b(u) \\ c(u) & d(u) \end{pmatrix} = \begin{pmatrix} a(w^2u + w) & \alpha^2 b(w^2u + w) \\ \alpha^{-2}c(w^2u + w) & d(w^2u + w) \end{pmatrix} \cdot \begin{pmatrix} a(-wu) & \alpha b(-wu) \\ \alpha^{-1}c(-wu) & d(-wu) \end{pmatrix}.$$
$$(35)$$

The solution was found numerically, and the coefficients of polynomial expansions for $a(u)$, $b(u)$, $c(u)$, $d(u)$ are listed in Ref. [12]. The factor α was also computed, so

$$\alpha = 2.890053525\ldots \quad \text{and} \quad \beta = -w^{-1} = 1.6180339\ldots \qquad (36)$$

These two constants determine scaling properties of the critical attractor on the (x, u)-plane. In fact, the variable x in the RG equation is not the same as in the original map: we need to introduce scaling coordinates in the (x, u)-plane. As found numerically [12], the variable change looks like

$$X \propto x + 2.34719526 + 5.92667u - 210.629u^2, \quad U = u. \qquad (37)$$

Figure 11 illustrates scaling property of the critical atttractor. Observe excellent reproduction of details of the structure in scaling coordinates (X, u).

Numerical solution of the eigenvalue problem (6) for the fractional-linear fixed point reveals two relevant eigenvalues

$$\delta_1 = 3.134272989\ldots \text{ and } \delta_2 = w^{-1} = 1.618033979\ldots \qquad (38)$$

They are responsible for scaling properties of the parameter space near the critical point. If we depart from the critical point along the bifurcation curve, the first eigenvector does not contribute; the relevant perturbations are associated with δ_2. If we choose a transversal direction, say, along the axis b, the perturbation of the first kind (δ_1) appears.

In the case under consideration we have $\delta_1 > \delta_2$ and $\delta_1 > \delta_2^2$, but $\delta_1 < \delta_2^3$, so only linear and quadratic terms must be taken into account in the parameter change. The scaling coordinates (C_1, C_2) are linked with parameters of the original map as

$$b = b_{TF} + C_1 - 0.64938C_2 - 0.33692C_2^2, \quad \varepsilon = 2 + C_2. \qquad (39)$$

To illustrate scaling associated with the nontrivial constant δ_1 let us consider duration of laminar phases in a course of intermittent dynamics generated by the map (33). In usual Pomeau – Manneville intermittency of type I the average duration of the laminar stages behaves as $\langle t_{\text{lam}} \rangle \propto \Delta b^{\nu}$ with $\nu = 0.5$ [23-26]. In presence of the quasiperiodic force the same law is valid in the subcritical region, $\varepsilon < 2$. In the critical case $\varepsilon = 2$ the exponent is distinct. Indeed, as follows from the RG results, to observe increase of a characteristic time scale by factor $\theta = w^{-1} = 1.61803$ we have to decrease a shift of parameter b from the bifurcation threshold by factor $\delta_1 = 3.13427$. As follows, the exponent must be $\nu = \log\theta / \log\delta_1 \cong 0.42123$. Figure 12 shows data of numerical experiments with the fractional-linear map. At each fixed ε in average duration of passage through the "channel" near the formerly existed attractor-repeller pair was computed in dependence on Δ for ensemble of orbits with random initial conditions. Results are plotted in the double logarithmic scale. For particular $\varepsilon = 1.7$ (subcritical) and 2 (critical) the dependencies fit the straight lines of a definite slope, estimated as 0.508 and 0.424, in good agreement with the theory. At subcritical ε slightly less than 2 one can observe a "crossover" phenomenon: the slope changes from the critical to the subcritical one at some intermediate value of Δb.

8. Conclusion

The present paper was devoted to a review of critical situations at the onset of chaotic or strange nonchaotic behavior via quasiperiodicity, more concretely, in the case of the golden-mean ratio of the basic frequencies. We

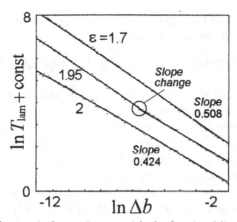

Figure 12. Data of numerical experiments with the fractional-linear map: average duration of passage through the 'channel' versus deflection from the bifurcation threshold for three values of ε in the double logarithmic scale. Observe a 'crossover' phenomenon, the slope change from critical to subcritical value at some intermediate value of Δb for $\varepsilon = 1.95$.

have derived a two-dimensional generalization of the Feigenbaum-Kadanoff-Shenker RG equation and demonstrate that it may be used to treat a number of critical situations – the conventional golden-mean criticality (GM) and the critical behaviors in quasiperiodically driven model maps: torus collision terminal (TCT), torus-doubling terminal (TDT), and torus fractalization at the intermittency threshold (TF). All these critical situations are of obvious interest for a problem of synchronization in nonlinear systems, in the context of study of transitions associated with break-up or other bifurcations of complex generalized synchronous regimes. In perspective, it would be interesting to reveal details and regularities of coexistence and subordination of all the types of critical behavior.

As is common in situations allowing the RG analysis, one can expect that the quantitative regularities intrinsic to our model maps will be valid also in other systems relating to the same universality classes. It would be significant to find these types of behavior in systems of higher dimension, for example, in quasiperiodically driven invertible 2D maps, which could represent Poincare maps of some flow systems. It would be interesting to arrange special experiments on search and observation of the considered types of critical behavior in physical sysyems. Since now, only GM critical behavior was studied experimentally in some details, see e.g. [35,20, 36], and the TDT criticality was observed in a particular quasiperiodically forced electronic system [33,34]).

Acknowledgements

I thank U. Feudel, E. Neumann, A. P. Kuznetsov, A. Pikovsky, and I. Sataev for fruitful collaboration, discussions, and valuable help during a work on different parts of the present research.

This work was supported by RFBR (grants No 00-02-17509 and 03-02-16074) and CRDF (award REC-006).

References

1. A. Pikovsky, M. Rosenblum, and J. Kurths. Synchronization: Universal Concept in Nonlinear Sciences. Cambridge University Press, 2001.
2. C.Grebogi, E.Ott, S.Pelikan and J.A.Yorke. Strange attractors that are not chaotic. Physica **D13**, 1984, 261-268.
3. F.J.Romeiras, A.Bondeson, E.Ott, T.M.Antonsen, and C.Grebogi. Quasiperiodically forced dynamical systems with strange nonchaotic attractors. Physica **D26**, 1987, 277-294.
4. A.S.Pikovsky and U.Feudel. Characterizing strange nonchaotic attractors. Chaos **5**, 1995, 253-260.
5. M.J.Feigenbaum. Quantitative universality for a class of nonlinear transformations. J. Statist. Phys. **19**, 1978, 25-52.
6. M.J.Feigenbaum. The universal metric properties of nonlinear transformations, J. Statist. Phys. **21**, 1979, 669-706.
7. M.J.Feigenbaum, L.P.Kadanoff, S.J.Shenker. Quasiperiodicity in dissipative systems. A renormalization group analysis. Physica **D5**, 1982, 370-386.
8. D.Rand, S.Ostlund, J.Sethna, and E.D.Siggia. A universal transition from quasiperiodicity to chaos in dissipative systems. Phys. Rev. Lett. **49**, 1082, 132-135.
9. S.Ostlund, D.Rand, J.Sethna, and E.D.Siggia. Universal properties of the transition from quasi-periodicity to chaos in dissipative systems. Physica **D8**, 1983, 303-342.
10. S.P.Kunetsov, U.Feudel and A.S.Pikovsky. Renormalization group for scaling at the torus-doubling terminal point. Phys.Rev.**E57**, 1998, 1585-1590.
11. S.P.Kuznetsov, E.Neumann, A.Pikovsky, I.R.Sataev. Critical point of tori-collision in quasiperiodically forced systems. Phys.Rev. **E62**, 2000, No 2, 1995-2007.
12. S.P.Kuznetsov. Torus fractalization and intermittency. Phys.Rev. **E65**, 2002, 066209.
13. S. J. Shenker. Scaling behavior in a map of a circle onto itself. Empirical results. Physica **D5**, 1982, 405-411.
14. K.Kaneko. Doubling of torus. Progr.Theor.Phys., **69**, 1983, 1806-1810.
15. S.P.Kuznetsov. Effect of a periodic external perturbation on a system which exhibits an order-chaos transition through period-doubling bifurcations. JETP Lett. **39**, 1984, No 3, 133-136.
16. K.Kaneko. Oscillation and doubling of torus. Progr.Theor.Phys. **72**, No 2, 1984, 202-215.
17. A.Arneodo. Scaling for a periodic forcing of a period-doubling system. Phys.Rev.Lett. **53**, 1984, 1240-1243.
18. S.P.Kuznetsov, A.S.Pikovsky. Renormalization group for the response function and spectrum of the period-doubling system. Phys.Lett. **A140**, 1989, 166-172.
19. A.Arneodo, P.H.Collet, E.A.Spiegel. Cascade of period doublings of tori. Phys.Lett. **A94**, 1983, 1-4.

20. V.S.Anishchenko. Dynamical Chaos - Models and Experiments. Appearance, Routes and Structure of Chaos in Simple Dynamical Systems (World Scientific, Singapore, 1995).

21. U.Feudel, A.S.Pikovsky, and J.Kurths. Strange non-chaotic attractor in a quasiperiodically forced circle map, Physica **D88**, 1995, 176-186.

22. H.Osinga, J.Wiersig, P.Glendinning and U.Feudel. Multistability and nonsmooth bifurcations in the quasiperiodically forced circle map. Int. J. Of Bifurcation and Chaos, **11**, 2001, 3085-3107.

23. Y.Pomeau, P.Manneville. Intermittent transition to turbulence in dissipative dynamical systems. Commun. Math. Phys. **74**, 1980, 189-197.

24. B.Hu, J.Rudnik. Exact solution of the Feigenbaum renormalization group equations for intermittency. Phys.Rev.Lett., **48**, 1982, 1645-1648.

25. J.E.Hirsch, B.A.Huberman, and D.J.Scalapino. Theory of intermittency. Phys. Rev. **A25**, 519-532 (1982).

26. F.Argoul and A.Arneodo. Scaling for periodic forcing at the onset of intermittency. J. Phys. Lett. (Paris) **46**, L901 (1985).

27. K.M.Briggs, T.W.Dixon, G.Szekeres. Analytic solution of the Cvitanovic-Feigenbaum and Feigenbaum-Kadanoff-Shenker equations. Int.J.of Bifurcation and Chaos, **8**, 1998, 347-357.

28. T.W.Dixon, T.Gherghetta, B.G.Kenny. Universality in the quasiperiodic route to chaos. CHAOS, **6**, 1996, 32-42.

29. N.Yu.Ivankov, S.P.Kuznetsov. Complex periodic orbits, renormalization and scaling for quasiperiodic golden-mean transition to chaos. Phys.Rev. **E63**, 2001, 046210.

30. S.P.Kuznetsov. Dynamical Chaos. Moscow, Fizmatlit, 2001, 296p. (In Russian.)

31. A.P.Kuznetsov, S.P.Kuznetsov, I.R.Sataev. A variety of period-doubling universality classes in multi-parameter analysis of transition to chaos. Physica **D109**, 1997, 91-112.

32. A.P.Kuznetsov, S.P.Kuznetsov, I.R.Sataev. Three-parameter scaling for one-dimensional maps. Phys.Lett **A189**, 1994, 367-373.

33. B.P.Bezruchko, S.P.Kuznetsov, A.S.Pikovsky, Ye.P.Seleznev, U.Feudel. On dynamics of nonlinear systems under external quasi-periodic force near the terminal point of the torus doubling bifurcation curve. Applied Nonlinear Dynamics, **5**, 1997, No 6, 3-20. (In Russian.) See also http://www.sgtnd.tserv.ru/eng/index.htm.

34. B.P.Bezruchko, S.P.Kuznetsov, Ye.P.Seleznev. Experimental observation of dynamics near the torus-doubling terminal critical point. Phys.Rev. **E62**, 2000, No 6, 7828-7830.

35. J.A.Glazier, G.Gunaratne, A.Libchaber. $F(\alpha)$ curves – Experimental results. Phys.Rev. **A37**, 1988, 523-530.

36. D.Barkley, A.Cumming. Thermodynamics of the quasi-periodic parameter set at the borderline of chaos – Experimental results. Phys.Rev.Lett., **64**, 1990, 327-331.

SYNCHRONIZATION AND CLUSTERING IN ENSEMBLES OF COUPLED CHAOTIC OSCILLATORS

YU. MAISTRENKO[1], O. POPOVYCH[1,2] and S. YANCHUK[1]
[1]*Institute of Mathematics, National Academy of Sciences of Ukraine, Tereshchenkivska St. 3, 01601 Kyiv, Ukraine*
[2]*Department of Physics, University of Potsdam, PF 601553, 14415 Potsdam, Germany*

Abstract

When identical chaotic oscillators interact, a state of complete or partial synchronization may be attained, providing a special kind of dynamical patterns called clusters. The simplest, coherent clusters arise when all oscillators display the same temporal behavior. Others, more complicated clusters are developed when population of the oscillators splits into subgroups such that all oscillators within a given group move in synchrony. Considering a system of mean-field coupled logistic maps, we study in details the transition from coherence to clustering and demonstrate that there are four different mechanisms of the desynchronization: riddling and blowout bifurcations, appearance of symmetric and asymmetric clusters. We also investigate the cluster-splitting bifurcation when the underlying dynamics is periodic. For the system of three and four coupled Rössler oscillators, we prove the existence of clusters and describe related bifurcations and in-cluster dynamics.

1. Introduction. Ensembles of globally coupled logistic maps

We study cluster formation phenomenon in the ensemble of N globally coupled chaotic maps (Sec. 1–6)

$$x_i(n+1) = (1-\varepsilon)f\left(x_i\left(n\right)\right) + \frac{\varepsilon}{N}\sum_{j=1}^{N} f\left(x_j\left(n\right)\right), \qquad (1)$$

101

A. Pikovsky and Y. Maistrenko (eds.), Synchronization: Theory and Application, 101–138.
© 2003 Kluwer Academic Publishers. Printed in the Netherlands.

where $i = 1, \ldots, N$ is a space index for the N-dimensional state vector $\mathbf{x} = \{x_i(n)\}_{i=1}^{N}$, and $n = 0, 1 \ldots$ is the discrete time variable. $\varepsilon \in \mathbb{R}$ is the coupling parameter and $f : \mathbb{R} \to \mathbb{R}$ is a one-dimensional noninvertible map for which we shall assume the form $f(x) = ax(1 - x)$ (the logistic map). a will be referred to as the nonlinearity parameter of the individual map, and the N-dimensional map system defined by Eq. (1) will be denoted Φ.

The simplest form of asymptotic dynamics that can arise in the globally coupled map system (1) is the fully synchronized (or coherent) state in which all elements display the same temporal variation. In this case the motion is restricted to a one-dimensional invariant manifold $D = \{(x_1, x_2, \ldots, x_N) \mid x_1 = x_2 = \ldots = x_N\}$, the main diagonal in N-dimensional phase space, and along this manifold the dynamics is governed by the one-dimensional map f of the individual oscillator. For certain values of the parameters a and ε, the coherent state may attract all trajectories starting from points in its N-dimensional neighborhood. In this case, the coherent state is asymptotically stable. For other values of ε, the phenomenon of clustering (or partial synchronization) may occur, i.e., the population of oscillators splits into subgroups (clusters) with different dynamics, but such that all oscillators within a given cluster asymptotically move in synchrony.

In his original work, Kaneko [1] developed a rough phase diagram for the occurrence of different clustering states in the globally coupled map system (1). If the coupling parameter ε is high enough (e.g., $\varepsilon > 0.355$ for $a = 3.8$), the state of full synchronization attracts almost all trajectories within a large region. For lower values of ε, the coherent state breaks up into a number of clusters. Immediately below the coherent state one typically finds an ordered state with two-cluster dynamics, or, for higher values of a, a so-called glassy phase where a few large clusters appear to coexist with many small clusters. Finally, as the coupling parameter becomes small enough, a transition to a turbulent state takes place. Here, almost all attracting states involve a large number of clusters, and the oscillators are nearly completely desynchronized.

In subsequent works, Kaneko has applied the globally coupled map approach as a model of biological cell differentiation [2]. He has also studied the occurrence of Milnor attractors and the role of noise-induced selection in high-dimensional systems [3]. Referring to the original definition [4], a Milnor attractor is a state that attracts a positive Lebesgue measure set of points from its neighborhood, but for which this neighborhood may also contain a positive Lebesgue measure set of points that are repelled from the (weakly) attracting state. The existence of such weak attractors is closely linked to the recently discovered phenomena of riddled basins of attraction [5, 6] and on-off intermittency [7].

Kaneko's work (see [8] for more complete list of references) has also inspired a considerable number of other investigators. In particular, the glassy state from the Kaneko's bifurcation diagram has been studied by Manrubia and Mikhailov [9] discussing a very long transient in globally coupled map systems. Manifestation of periodicity in the so-called turbulent regime has been considered by Shimada and Kikuchi [10] who showed how the most symmetric three-cluster attractor with period-3 motion is related to the period-3 window of the individual map. An extensive bifurcation analysis of the loss of coherence and emergence of two-cluster dynamics has been performed by Balmforth et al., [11]. Xie et al. [12] have discussed the transverse destabilization of synchronous periodic orbits in the period-doubling cascade in the main periodic windows of coupled subsystems and a particular case of the cluster-splitting cascade, cluster doubling, has been demonstrated for a system of globally coupled logistic maps, systems of globally coupled Duffing oscillators, and Josephson junction series arrays. Glendinning [13] has investigated the fractal nature of the blowout bifurcation in which the coherent state loses its average stability in the transverse direction, illustrating how globally coupled map systems can proceed through a complicated sequence of synchronizations and desynchronizations in connection with transitions between periodic and chaotic dynamics for the individual map.

Interesting phenomenon of non-trivial collective dynamics in coupled systems is a subject of intensive study with use their transfer (Frobenius-Perron) operator [14], finite-size collective Lyapunov exponents [15, 16], directed percolation universality class [17], a linear response function [18]. Asymptotic behavior and statistical properties of a globally coupled map system in the thermodynamic limit $N \to \infty$, has been considered, in particular, by Pikovsky and Kurths [19] and by Hamm [20]. Some other aspects of synchronized and clustered dynamics of globally coupled systems have been discussed, e.g., in Refs. [21, 22, 23, 24], also for models with both local and global coupling [25, 26, 27] as a way of understanding hierarchical pattern formation in systems with interactions on different length scales. In this connection it is worth noticing that globally coupled systems differ qualitatively from locally coupled systems with respect to the types of dynamics that they can support. At the numerical study of the coupled systems, computer calculations recover the spurious phenomenon known as synchronization with positive conditional (i.e. transverse) Lyapunov exponents [28]. This type of numerically generated clustering can lead to false conclusions concerning the occurrence of low-dimensional dynamics because of a *finite precision phenomenon* [29] which can be avoided with the use of special numerical technics proposed by Pikovsky et al. [30] or by explicitly examining stability of obtained clustered states.

Theoretical study of the synchronized and clustered dynamics of globally coupled systems has recently attained an experimental support. Wang et al. [31] have provided experimental evidence of clustering in a system of globally coupled electrochemical reactors. Clustering has been also observed in some other globally coupled and spatially extended systems [32, 33, 34]. Important questions that arise in this connection relate to the types of clustering states that can be realized and to the bifurcations through which the transitions from the coherent behavior to cluster dynamics take place.

Considering a system of two coupled, identical logistic maps, Maistrenko et al. [35, 36] have performed a detailed investigation of the riddling bifurcation [37, 38] in which the first transverse destabilization of a periodic orbit embedded in the synchronized chaotic state takes place. The role of absorbing areas [39] in the riddling transitions for coupled non-invertible maps has been emphasized [40]. The same model can be used to study the transition to two-cluster dynamics for a system of N coupled logistic maps, provided that the maps distribute themselves symmetrically between the two clusters. In a subsequent study by Popovych et al. [41], emphasis was given to the role of an asymmetric distribution of the oscillators. Whereas the transverse period-doubling bifurcation remains essentially unaffected by such an asymmetry, the transverse pitchfork bifurcation was found to be replaced by a transcritical riddling bifurcation in which a periodic orbit born in a saddle-node bifurcation passes through the synchronization manifold and exchanges its transverse stability with a saddle cycle of similar periodicity in that manifold.

The phenomenon of chaotic partial synchronization (or cluster formation) has been studied by Hasler et al. [42] for a system of three coupled skew-tent maps. Applying a special coupling scheme of relevance in connection with applications for secure communication, they have determined the regions in parameter space where total and partial synchronization take place and they have analyzed the bifurcations through which the coherent state (total synchronization) breaks down to give way for two- and three-cluster dynamics. It was shown that chaotic cluster states arising after blowout bifurcation of a coherent attractor cannot be asymptotically stable in the whole N-dimensional phase space of the globally coupled map system. For a system of N globally coupled logistic maps, Popovych et al. [43, 44] have given an extended description of loss of stability by chaotic coherent state and emergence of two-cluster states. By demonstrating that both symmetric and asymmetric two-cluster states are important for cluster formation process they refined the Kaneko's bifurcation diagram [1]. In the case, where the underlying dynamics of logistic map is periodic, the considered system exhibits cascades of cluster-splitting bifurcations when coupling parameter varies, which has been studied in details in Ref. [45].

The simplest clustered dynamics is characterized by a behavior in which two synchronized groups of oscillators (two clusters) are present:

$$x_1 = x_2 = \ldots = x_{N_1} \overset{def}{=} x$$
$$x_{N_1+1} = x_{N_1+2} = \ldots = x_N \overset{def}{=} y, \tag{2}$$

where $N_1 < N$ and $N_2 = N - N_1$ denote the number of synchronized elements in each of the two clusters.

Under these conditions the N-dimensional coupled map system (1) reduces to a system of two coupled one-dimensional maps

$$\begin{aligned} x(n+1) &= f(x(n)) + p\varepsilon[f(y(n)) - f(x(n))], \\ y(n+1) &= f(y(n)) + (1-p)\varepsilon[f(x(n)) - f(y(n))], \end{aligned} \tag{3}$$

where the cluster asymmetry parameter p describes the relative distribution of oscillators between the two clusters. More precisely, $p = N_2/N$ denotes the fraction of the total population that synchronizes in state y. For $N = 3$, for instance, with two clusters $x_1 = x_2 \overset{def}{=} x$ and $x_3 \overset{def}{=} y$, $N_1 = 2$, $N_2 = 1$ and the dynamics is described by system (3) with $p = 1/3$. Clearly, for $N = 3$, two-cluster dynamics can be realized in $3!/(2!1!) = 3$ different ways. Hence, we have three distinct (and mutually symmetric) two-cluster states. For larger values of N, the possible realizations of a given cluster distribution grow very rapidly.

Transverse stability of the two-cluster state of the form (2), i.e., stability of the state in the direction transverse to the clustered subspace defined by (2), is determined by the transverse Lyapunov exponents for the considered two-cluster states. As it follows from the symmetry of (1), for any two-cluster state $A^{(2)}$ and for any dimension N, there are only two distinct transverse Lyapunov exponents [43, 46]

$$\lambda_{\perp,1}^{(2)} = \lim_{k \to \infty} \frac{1}{k} \sum_{n=0}^{k-1} \ln | f'(x(n))(1 - \varepsilon) |$$

$$\tag{4}$$

$$\lambda_{\perp,2}^{(2)} = \lim_{k \to \infty} \frac{1}{k} \sum_{n=0}^{k-1} \ln | f'(y(n))(1 - \varepsilon) |$$

evaluated for a typical trajectory $\{(x(n), y(n))\}_{n=0}^{\infty} \subset A^{(2)}$. The first exponent $\lambda_{\perp,1}^{(2)}$ is responsible for the breakdown of the clustered state labelled by x in (2), whereas $\lambda_{\perp,2}^{(2)}$ is responsible for the breakdown of the clustered state y.

Consider the case where the state vector $\mathbf{x} = \{x_i\}_{i=1}^N$ of the N-dimensional system (1) splits into K groups such that in each group the coordinates become identical:

$$x_{i_1} = x_{i_2} = \ldots = x_{i_{N_1}} \overset{def}{=} y_1$$
$$x_{i_{N_1+1}} = x_{i_{N_1+2}} = \ldots = x_{i_{N_1+N_2}} \overset{def}{=} y_2 \qquad (5)$$
$$\cdots\cdots\cdots\cdots\cdots\cdots\cdots\cdots\cdots\cdots$$
$$x_{i_{N_1+N_2+\ldots+N_{K-1}+1}} = x_{i_{N_1+N_2+\ldots+N_{K-1}+2}} = \ldots = x_{i_N} \overset{def}{=} y_K.$$

The positive integer N_j indicates the number of identical variables x_i belonging to the jth cluster, $j = 1, \ldots, K$, so that $N_1 + N_2 + \ldots + N_K = N$. We note that, by virtue of the complete symmetry of the system (i.e., the fact that all the individual maps are the same), for any set $\{N_j\}$ the K-dimensional subspace defined by Eqs. (5) remains invariant for the dynamics in the corresponding K-cluster state.

Introducing the set of the cluster asymmetry parameters $p_j = N_j/N, j = 1, \ldots, K$, the dynamics in the K-cluster state can be described by the system of K coupled one-dimensional maps

$$y_i(n+1) = (1-\varepsilon)f(y_i(n)) + \varepsilon \sum_{j=1}^{K} p_j f(y_j(n)), \quad i = 1, \ldots, K. \qquad (6)$$

This system is also a globally coupled map system, but with different weights p_j associated with the contribution of the jth cluster to the global coupling. Varying the parameters p_j in (6) one can obtain the governing map for any possible K-cluster dynamics of the original system (1).

A necessary condition for the presence of stable K-cluster behavior in system (1) is that the map (6) with the assumed values of the parameters p_j has a stable invariant set $A^{(K)}$, but that there is no stable invariant sets $A^{(L)} \supseteq A^{(K)}$ with $L < K$. For example, system (1) with even number of cites N may demonstrate symmetric two-cluster dynamics (2) if the two-dimensional map (3) with $p = 0.5$ has a stable invariant set $A^{(2)}$ which does not belong to the diagonal $D = \{(x,y) \mid x = y\}$.

Provided that it is stable in the cluster subspace, the conditions for an attractor $A^{(K)}$ of system (6) to be stable in the whole N-dimensional phase space are that it is also stable in the transverse directions. The transverse stability of $A^{(K)}$ may be asymptotic, when it attracts all trajectories from its N-dimensional neighborhood $\mathcal{U}^{(N)}$, or weak, when $A^{(K)}$ is stable in the Milnor sense, i.e., it attracts a positive Lebesgue measure set of initial data from $\mathcal{U}^{(N)}$ [4]. To determine K transverse Lyapunov exponents $\lambda_{\perp,j}$ for a K-cluster state (5) possessing the attractor $A^{(K)}$, one can iterate the map

(6) on $A^{(K)}$ and calculate them by the formula

$$\lambda^{(K)}_{\perp,j} = \lim_{k \to \infty} \frac{1}{k} \sum_{n=0}^{k-1} \ln | f'(y_j(n)) | + \ln |1 - \varepsilon|, \quad j = 1, 2, \ldots, K. \quad (7)$$

When all the Lyapunov exponents are negative, $A^{(K)}$ is an attractor in N dimensions in the Milnor sense [4]. This provides the conditions for the existence of stable K-cluster states for system (1).

2. On the way from coherence to clustering

We find that the transition from coherence to clustering in the globally coupled system (1) is ruled by four different mechanisms:

 — *riddling* bifurcation;
 — *blowout* bifurcation;
 — appearance of *symmetric clusters*;
 — appearance of *asymmetric clusters*.

Phase diagram for the bifurcations is shown in Fig. 1 which presents a two-dimensional bifurcation diagram in the (a, ε)-parameter plane for the desynchronization of the coherent motion and for the emergence of two-cluster states in the system (1). Figure 1 can be compared directly with the rough phase diagram provided by Kaneko [1]. The fully drawn noisy (fractal) curve represents the blowout bifurcation of the synchronized coherent state. Below this curve the coherent motion is repelling on average. The dotted curve denoted "riddling" represents the transverse destabilization of the symmetric nontrivial fixed point (x_0, x_0). The solid curve with the characteristic step-like structure represents the emergence of highly asymmetric two-cluster states in system (1). The numbers associated with this curve denote the periods of stable in-cluster cycles that cause the emergence of asymmetric two-cluster states. The two remaining bifurcation curves in Fig. 1 are for the stabilization of the symmetric $(p = \frac{1}{2})$ two-cluster states with period-2 (thin solid curve) and period-4 (thin dashed curve) in-cluster dynamics, respectively.

When the coupling strength ε starts to decrease, first the chaotic attractor $A^{(s)} \subset D$ loses its asymptotic transverse stability in a *riddling* bifurcation [37, 38]. This occurs when the first trajectory embedded in the synchronous chaotic state becomes transversely unstable. After the riddling bifurcation, $A^{(s)}$ is no longer stable in the Lyapunov sense. In any small neighborhood of the attractor one can find a positive measure set of phase points such that the trajectories, when starting from these points, will go away from $A^{(s)}$. Provided that other asymptotic states, which can be

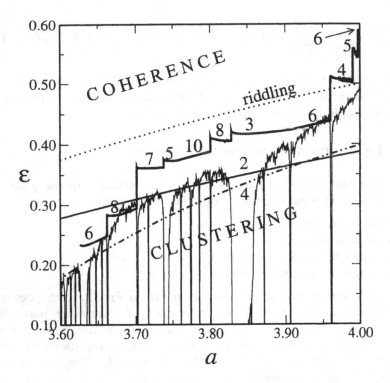

Figure 1. Phase diagram for cluster formation in a system of globally coupled logistic maps (1). a is the nonlinearity parameter for the individual map, and ε is the coupling parameter. The uppermost (dotted) curve represents the riddling bifurcation of the one-piece chaotic coherent state $A^{(s)}$ in which the fixed point $P_1^{(s)} \in A^{(s)}$ loses its transverse stability, and the fully drawn fractal curve delineates the blowout bifurcation. The smooth fully drawn and dashed bold curves represent stabilization of the asynchronous period-2 and period-4 cycles giving birth to the symmetric two-cluster states, respectively. The solid bold curve with the characteristic step-like structure represents the emergence of highly asymmetric two-cluster states in system (1), and the numbers associated with this curve denote the periods of stable in-cluster cycles that cause the emergence of asymmetric two-cluster states.

reached from the neighborhood of $A^{(s)}$, do not exist, most of the trajectories will sooner or later return to the neighborhood of $A^{(s)}$. In the presence of noise, some of the trajectories may again perform a burst, manifesting the typical bubbling behavior [37]. This type of characteristic phase dynamics is associated with the Milnor stability of $A^{(s)}$ [4]. It gives rise to locally riddled basins of attraction for the synchronous chaotic state [37, 38].

In the phase diagram of Fig. 1, the uppermost (dotted) curve denotes the transverse destabilization of the fixed point $P_1^{(s)} = (x_0, x_0, \ldots, x_0), x_0 = 1 - \frac{1}{a}$. In the parameter-$a$ regime, where synchronous attractor $A^{(s)}$ is one-piece chaotic ($a > a_0 \cong 3.678573$), $P_1^{(s)}$ is the first trajectory on $A^{(s)}$ to lose its transverse stability, and, hence, the dotted curve represents the riddling bifurcation curve. This curve can easily be determined analytically [35]. Destabilization of $P_1^{(s)}$ takes place via a transverse period-doubling bifurcation and produces an asynchronous period-2 saddle around the fixed point. For slightly lower values of the coupling parameter ε, the synchronous period-2 cycle embedded in the coherent chaotic state also undergoes a transverse period-doubling, producing an asynchronous period-4 saddle.

The fractal curve in Fig. 1 denotes the *blowout* bifurcation of $A^{(s)}$. The blowout occurs at $\varepsilon = \varepsilon_{bl} = 1 - e^{-\lambda_a}$ when the transverse Lyapunov exponent $\lambda_\perp^{(1)}$ of the synchronous chaotic set changes its sign from minus to plus. After the blowout bifurcation, $A^{(s)}$ is no longer an attractor but has turned into a chaotic saddle. Almost all trajectories now go away from the coherent state described by the chaotic set $A^{(s)}$, and in general only a zero measure set of trajectories will approach $A^{(s)}$ [37]. One of the main questions of the present paper is to determine the fate of the diverging trajectories. We find that, depending sensitively on a, there are different possibilities associated with the mutual disposition of the blowout and cluster stabilization curves as well as with system dimension N.

Let a be fixed and let us consider what happens as the control parameter ε is reduced. If the blowout bifurcation occurs before the appearance of stable two-cluster states, the coherent phase turns into a high-dimensional chaotic state. With further reduction of parameter ε, this may be captured into one of the periodic two-cluster states. In the opposite situation, i.e. when the asynchronous periodic cycles stabilize before the blowout bifurcation, two-cluster states appear before the blowout of the coherent state. As a consequence, both types of dynamics − fully synchronized chaotic and two-cluster periodic − coexist in some region of the (a, ε)-parameter plane [41].

In Fig. 1, the solid and dashed bold curves represent the stabilization of the asynchronous cycles P_2 (period-2) and P_4 (period-4) forming the possible symmetric (or close to symmetric) two-cluster states. These cycles remain stable in some regions under the curves to destabilize with further reduction of ε in a Hopf bifurcation. The symmetric two-cluster state P_2, which arises as the asynchronous saddle cycle produced through a transverse period-doubling bifurcation of the symmetric fixed point $P_1^{(s)}$, stabilizes in a subcritical, inverse pitchfork bifurcation along the fully drawn bold curve. P_4, which arises from a transverse period-doubling of the symmetric period-2 orbit, stabilizes along the dashed bold curve. It can be seen in Fig. 1 that, for $a \gtrsim 3.93$, P_4 stabilizes before (i.e. for higher values of ε than) P_2. Moreover, slightly asymmetric two-cluster states stabilize after the symmetric ones when ε decreases.

3. Beyond the blowout bifurcation: transverse instability of chaotic clusters

In this section we show that the chaotic motions in the two- and three-cluster states often observed after the blowout bifurcation, in general, are transversely unstable. To verify this, we show that the largest transverse Lyapunov exponents $\lambda_\perp^{(2)}$ and $\lambda_\perp^{(3)}$ for the two- and three-cluster states are positive. Moreover, immediately after the blowout bifurcation at $\varepsilon = \varepsilon_{bl}$ they grow in accordance with a power law.

3.1. CHAOTIC TWO-CLUSTER STATE

Figure 2a displays a scan of $\lambda_\perp^{(2)}$ over the range from $\varepsilon = 0$ (uncoupled system) to right above the blowout bifurcation ($\varepsilon_{bl} = 0.5$) for $a = 4$ and for the asymmetry parameter $p = 0.5$ (symmetric clusters). The scan of $\lambda_\perp^{(2)}$ is depicted by the bold curve. The dashed curves show the variation of the two Lyapunov exponents $\lambda_{\parallel,\{1,2\}}^{(2)}$ that control the two-dimensional cluster dynamics. In Fig. 2a there is an interval around $\varepsilon = 0.23$ where $\lambda_\perp^{(2)}$ is negative while the Lyapunov exponents in the two-cluster plane are positive. Here, we have a transversely stable chaotic two-cluster state. However, through most of the scan the transverse Lyapunov exponent is positive when the longitudinal exponents are positive. Note that, when ε decreases, the state stabilizes in N dimensions if it becomes an attracting cycle. Our interest is focused on the behavior immediately after the blowout bifurcation of the coherent state, that occurs at $\varepsilon_{bl} = 0.5$.

Figure 2b shows an enlargement of the rightmost parts of the graphs from Fig. 2a in order to illustrate the power law of growth for $\lambda_\perp^{(2)}$ of the chaotic two-cluster state $A^{(2)}$ arising immediately after the blowout bifurcation of the coherent state $A^{(s)}$ for three fixed values of the asymmetry parameter p. In all cases, the transverse Lyapunov exponent is positive (although small). Here, $a = 4$. As we can conclude

$$\lambda_\perp^{(2)} \sim |\varepsilon_{bl} - \varepsilon|^\alpha, \quad \varepsilon \to \varepsilon_{bl}, \tag{8}$$

where the exponent $\alpha \cong 2$ for the symmetric clusters ($p = 0.5$) and decreases for assymetric clusters. We conclude that the chaotic two-cluster

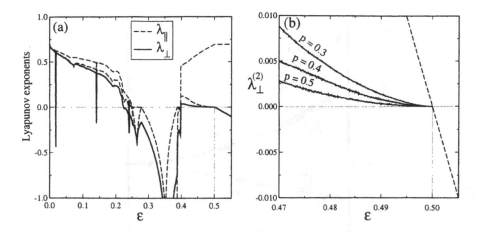

Figure 2. (a) Variation of the largest transverse Lyapunov exponent $\lambda_\perp^{(2)}$ (solid bold curve) with the coupling parameter ε for the two-cluster states being symmetric ($p = 0.5$). Parameter $a = 4$. Dashed curves represent in-cluster Lyapunov exponents $\lambda_\parallel^{(2)}$ within the two-cluster state. (b) Enlarged part from (a) around $\varepsilon = 0.5$. The three curves represent different values of the asymmetry parameter p. The dashed curve gives a variation of the transverse Lyapunov exponent $\lambda_\perp^{(1)}$ of the coherent state $A^{(s)}$.

state formed after blowout bifurcation of the chaotic coherent state cannot be stable in N-dimensional phase space.

3.2. CHAOTIC THREE-CLUSTER STATE

Since chaotic two-cluster motions, which appear after blowout bifurcation, are transversely unstable, it follows that dimension of the chaotic motions that arise must be larger than two. We now give a numerical evidence that the dimension must also be larger than three. For this, we show that chaotic motions in the symmetric three-cluster states are also transversely unstable.

Figure 3 presents a plot of the transverse Lyapunov exponent $\lambda_\perp^{(3)}$ versus coupling parameter ε for a symmetric three-cluster state. As one can see, $\lambda_\perp^{(3)}$ becomes positive immediately after the blowout bifurcation ($\varepsilon_{bl} = 0.5$) and appears to grow in accordance with a power law similar to (8). This can be justified as follows. As illustrated in Fig. 4, the typical trajectory in the chaotic three-cluster state spends most of the time very near the diagonal two-dimensional planes $\sigma_z = \{x = y, z\}$, $\sigma_y = \{x = z, y\}$, and $\sigma_x = \{x, y = z\}$. Moreover, it switches between these planes in an apparently random manner. From this observation we conclude that an approximate value for the transverse (to the three-cluster state) Lyapunov

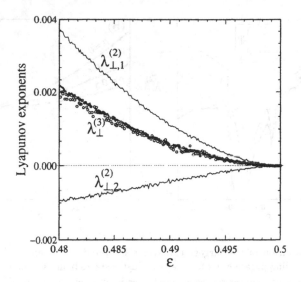

Figure 3.   Transverse Lyapunov exponent $\lambda_\perp^{(3)}$ (shown by circles) for a symmetric three-cluster state as a function of the coupling parameter ε. The largest $\lambda_{\perp,1}^{(2)}$ and the second $\lambda_{\perp,2}^{(2)}$ transverse Lyapunov exponents for the two-cluster state with $2:1$ $(p = 1/3)$ variable distribution between clusters are also shown. The value $\left(2\lambda_{\perp,1}^{(2)} + \lambda_{\perp,2}^{(2)}\right)/3$ is represented by bold dashed curve which fits the values of $\lambda_\perp^{(3)}$. Parameter $a = 4$.

exponent $\lambda_\perp^{(3)}$ can be obtained as calculated on the planes σ_x, σ_y, and σ_z, with the additional assumption that the average time spent near each of these planes is the same. This gives

$$\lambda_\perp^{(3)} \cong \left(2\lambda_{\perp,1}^{(2)} + \lambda_{\perp,2}^{(2)}\right)/3, \tag{9}$$

where $\lambda_{\perp,1}^{(2)}$ and $\lambda_{\perp,2}^{(2)}$ are the largest and the second transverse Lyapunov exponents for the chaotic motions in the two-cluster planes σ_x, σ_y, and σ_z. Using the expression (7) for the transverse Lyapunov exponents for two-cluster states and for three-cluster states ($K = 3$), we come to the approximate formula (9).

We note here that the numerical calculation of $\lambda_\perp^{(3)}$ has required the introduction of small noise of the order of 10^{-22}. Without the noise, trajectories are captured by the two-cluster dynamics in spite of the fact that, as we just shown, the chaotic two-cluster attractors are transversally unstable. It happens because of final precision in the calculations [30]. The average capturing times are shown in Fig. 5 for single (10^{-8}), double (10^{-16}) and triple (10^{-24}) precisions, respectively. We suppose that this capturing phe-

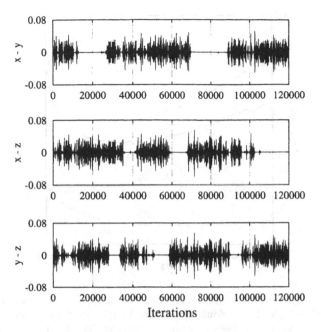

Figure 4.   Synchronization errors calculated on a typical trajectory for the chaotic three-cluster state (considering system (6) with $K = 3$ and $p_j = 1/3$, $j = 1, 2, 3$). We have added a small noise of the maximal amplitude 10^{-22}. The first 10^4 iterations are skipped, and the next $1.2 \cdot 10^5$ iterations are plotted. The trajectory spends most of its time near the two-dimensional planes $\sigma_z = \{x = y, z\}$, $\sigma_y = \{x = z, y\}$ and $\sigma_x = \{x, y = z\}$, and it switches between these planes in an apparently random manner. Parameters $a = 4$ and $\varepsilon = 0.495$.

nomenon can explain why high-dimensional chaotic motions arising after blowout bifurcation of the chaotic coherent phase have not previously been reported. Indeed, any regular calculation (without noise) gives evidence of two-cluster dynamics even though this is actually transversely unstable.

4.   Emergence of symmetric two-clusters

The appearance of the symmetric (or slightly asymmetric) two-cluster dynamics in the globally coupled map system (1) is caused by the stabilization of the period-2 or period-4 asynchronous cycles P_2 and P_4 of the two-dimensional system (3). In this section we consider how the moments of stabilization depend on a small cluster asymmetry, i.e., when the parameter p in system (3) starts to differ from 0.5. A main conclusion is that the symmetric clusters, i.e. with $p = 0.5$, stabilize before slightly asymmetric

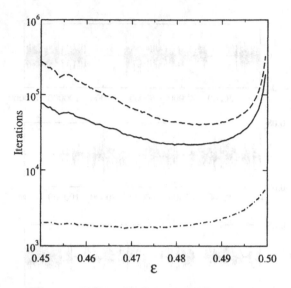

Figure 5. Average (over 8000 initial conditions) capturing time in a two-cluster state as calculated with single (10^{-8}), double (10^{-16}) and triple (10^{-24}) precisions and shown by dotted-dashed, solid, and dashed curves, respectively. By iterating system (6) with $K = 3$, $p_j = 1/3$, $j = 1, 2, 3$, and $a = 4$, we find that all trajectories are captured by transversally unstable two-cluster states in a finite time. The capturing phenomenon and the associated spurious stability with positive transverse Lyapunov exponents can be avoided by adding a small amount of noise to the numerical computations.

clusters. Moreover, the stabilization occurs the later the larger the asymmetry is. For the symmetric two-cluster state, the cycles P_2 and P_4 are born in transverse period-doubling bifurcations of the coherent fixed point $P_1^{(s)}$ and $P_2^{(s)}$, respectively. After the bifurcations they are first unstable (saddles) to later stabilize in inverse subcritical pitchfork bifurcations. A characteristic phase portrait of the in-cluster system (3) for the situation when both cycles P_2 and P_4 have already become stable is presented in Fig. 6.

For the case of slightly asymmetric clusters, the cycles P_2 and P_4 can be obtained by continuation of those in the symmetric case with the parameter p (starting with $p = 0.5$). If $p \neq 0.5$, these cycles stabilize in saddle-node bifurcations off the main diagonal rather than via inverse, subcritical pitchfork bifurcations as in the symmetric case $(p = 0.5)$.

Figure 7 shows the regions of stability for the various types of dynamics that evolve from P_2 and P_4 under variation of p and ε for two different values of the nonlinearity parameter a. In Fig. 7a $(a = 3.8)$, the upper boundary of the stability region (solid curve denoted SN) defines the moment of

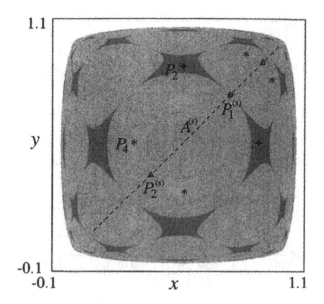

Figure 6. Phase portrait of the globally coupled map system (1) reduced to the symmetric two-cluster subspace ($p = 0.5$ in (3)) after stabilization of both the asynchronous period-2 (denoted by P_2 and plotted by crossed circles) and the asynchronous period-4 (denoted by P_4 and plotted by stars) cycles. After the blowout bifurcation, the coherent state $A^{(s)}$ (dashed line segment) is a chaotic saddle. The symmetric fixed point (denoted by $P_1^{(s)}$ and plotted by crossed square) and symmetric period-2 cycle (denoted by $P_2^{(s)}$ and plotted by triangles) are repellors being after the transverse period-doubling bifurcations which give birth to P_2 and P_4. Basins of attraction for the cycles P_2 and P_4 are shown in dark and light grey, respectively. Parameters $a = 3.9$ and $\varepsilon = 0.345$. With further reduction of ε each of the cycles P_2 and P_4 undergo a sequence of additional bifurcations leading to various forms of quasiperiodic and chaotic two-cluster dynamics.

stabilization of the asynchronous period-2 cycles P_2 in the afore-said saddle-node bifurcations. This curve is clearly seen to assume its maximal value for $p = 0.5$, representing the fact that symmetric clusters will stabilize before slightly asymmetric clusters as ε is reduced.

For $a = 3.8$, stabilization of P_4 occurs at lower values of the coupling parameter than stabilization of P_2, and we find the stability region for P_4 (and for solutions developed from P_4) in the upper right corner of the stability region for P_2. For $a = 4.0$ (Fig. 7b), on the other hand, P_4 stabilizes before P_2 (see Fig. 1), and the stability region for P_4 falls above that of P_2.

The stability of a periodic cycle in the two-cluster phase plane implies its stability in the whole N-dimensional phase space. Hence, the uppermost curves in Figs. 4a and b are the bifurcation curves in the (p, ε) -parameter plane for the appearance of the symmetric (or nearly symmetric) two-

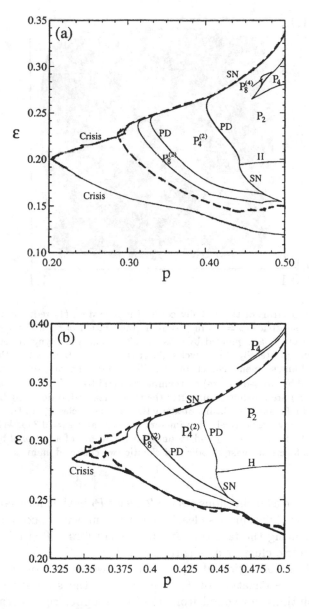

Figure 7. Stability regions in the (p, ε)-parameter plane for the various types of dynamics in system (3) that develop from the asynchronous period-2 (P_2) and period-4 (P_4) cycles and represent two-cluster states in (1). Bifurcation curves denoted by SN, PD, and H correspond to saddle-node, period-doubling, and Hopf bifurcations, respectively. With decreasing values of p we can follow P_2 through a cascade of period-doubling bifurcations into a chaotic off-diagonal attractor that finally destroys in a boundary crisis. Bold dashed curve bounds the region where the largest Lyapunov exponent transverse to the two-cluster state is negative. Here, system (1) displays stable two-cluster states with a distribution between clusters as defined by p and a dynamics that is given by the attractors developed from P_2. Parameters $a = 3.8$ in (a) and $a = 4$ in (b).

cluster states. The overlapping stability regions for P_2 and P_4 implies that the system has two coexisting types of two-cluster dynamics (see Fig. 6). With further variation of the parameters p and ε, the cycles P_2 and P_4 undergo a variety of different bifurcations in which more complicated two-cluster dynamics arises. Besides periodic cycles of higher periodicity, quasiperiodic and chaotic dynamics occur. Some of the bifurcation curves are indicated in Figs. 7 a and b where period-doubling and Hopf bifurcation curves are denoted PD and H, respectively.

If the attractor in a two-cluster state is quasiperiodic or chaotic, its stability within the two-cluster state does not imply its stability in the full N-dimensional phase space. The bold dashed curves in Figs. 7a and b denote the transverse destabilization of the two-cluster attractors developed from P_2, and the lower right curves represent their final boundary crises. The upper branch of the dashed curve coincides with the saddle-node bifurcation curve of two-cluster stabilization. As we can see, there is a fairly large parameter region where the attractor in the two-cluster state is quasiperiodic and yet transversely stable. Below this region there is another region where the two-cluster state is transversely unstable. More detailed study of the symmetric two-cluster states can be found in Ref. [43].

5. Emergence of asymmetric two-clusters

System (1) can demonstrate two-cluster behavior if the two-dimensional map F given by system (3) has an attractor $A^{(2)}$ that does not belong to the diagonal $D_2 = \{(x, y) \mid x = y\}$. As we have just seen, such an attractor usually originates in the stabilization of a periodic cycle out of the diagonal. Under variation of the parameters this stable cycle undergoes different types of bifurcations that may lead to other stable periodic cycles (e.g., via a period-doubling bifurcation) or to a stable closed invariant curve (via Hopf bifurcation). Further developments of the attractor $A^{(2)}$ may turn it into an attracting chaotic set which finally disappears in a boundary crisis. Therefore, to investigate the emergence of clusters in the system of coupled maps (1) we start by examining the appearance of stable point cycles within the appropriate subspace and then using the formulas (4) to verify the stability of these cycles with respect to perturbations perpendicular to the cluster subspace. More specifically, for two-cluster behavior, stable periodic cycles of the two-dimensional map (3) should be found by varying three parameters a, ε, and p.

For two fixed values of the nonlinearity parameter a, Fig. 8a,b display the regions in (p, ε)-parameter plane (shown in black), where the map (3) has an attracting cycle away from the diagonal D_2. Cycle periods are indicated by numbers. Hence, if the (p, ε)-parameter point falls in one of the

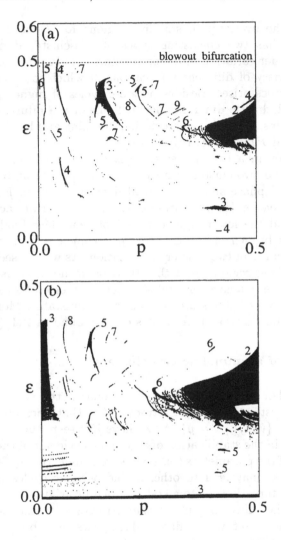

Figure 8.   Regions of (p, ε)-parameter plane where the map (3) displays an attracting periodic cycle outside of the diagonal. (a) $a = 4$,(b) $a = 3.84$, and (c) a = 3.6785735104. The moment of blowout bifurcation for the coherent chaotic state is shown by the dashed line in (a) and (c). For $a = 3.84$ regions of period-3 two-cluster dynamics are being formed both along the $\varepsilon = 0$ axis and along the $p = 0$ axis (b).

black regions of Fig. 8, system (1) can exhibit periodic two-cluster behavior for the corresponding value of parameter a. As discussed above, the only conditions for this two-cluster behavior of system (3) to be stable in the whole phase space of system (1) is that the corresponding in-cluster state is transversely stable. The distribution of the oscillators between the clusters is given by the value of p. To obtain the bifurcation diagrams in Fig. 8, we fixed parameter a, took a fine grid in the (p, ε)-parameter plane and, with 20 randomly chosen initial conditions for each grid point, iterated the map F to look for asymmetric stable cycles of a period up to 50. When such a cycle was found (at least for one initial condition), the corresponding (p, ε)-parameter point was plotted in black.

Inspection of Fig. 8 suggests that the first two-cluster states to appear as the coupling parameter ε is decreased from the coherent phase are highly asymmetric with respect to the distribution of oscillators between the clusters. Figure 8 also displays a surprising organization of the periodic regions to the right of the $p = 0$ value: They follow the well-known sequence of periodic windows for the logistic map. Indeed, as one can see in Figs. 8a and b the widest window corresponds to asymmetric two-cluster dynamics with period-3 behavior. The next, relatively large window is of period 5 followed by period 7 and 9. In between the period-3 and -5 windows there is a window of period 8.

To the left of the period-3 window we find a period-adding sequence of windows of periods 4, 5, 6 and so on (Fig. 8a, $a = 4$). These stability regions correspond to stable cycles $\gamma_k = \{x_i\}_{i=1}^k$, $k = 4, 5, \ldots$ of so-called maximal type that arise in the bifurcation diagram for the logistic map f_a beyond the period-3 window. For $a = 3.84$ (which is inside the period-3 window) such cycles have not yet appeared for the logistic map and, as it can be seen in Fig. 8b, the corresponding windows are not present in the (p, ε)-parameter plane of the map (3).

As follows, each time a periodic window under variation of the nonlinearity parameter a arises in the logistic map, the corresponding strongly asymmetric two-cluster state emerges from the $p = 0$ axis in the globally coupled map system. Moreover, the sequence of bifurcations that occur in conjunction with the periodic windows in the logistic map is recovered in the cluster formation process. Thus, immediately to the right of the period-3 two-cluster state we can observe a chaotic dynamics in the form of type-I intermittency, and to the left of the period-3 two-cluster state we find two-cluster states with period-6, period-12, etc., dynamics. Finally, it is interesting to note that the windows with two-cluster dynamics tend to reach to higher and higher values of ε the smaller p is. This implies that the first two-cluster states to synchronize when decreasing the coupling strength ε are those with a strongly asymmetric distribution.

In Fig. 9 the structure of the period-3 window is presented for $a = 4$. The stability window has now moved up to $p \approx 0.15$. The two-cluster attractor $A^{(2)}$ originating from the asymmetric period-3 cycle $P_3^{(a)}$ is not necessary stable in the whole N-dimensional phase space of the original ensemble

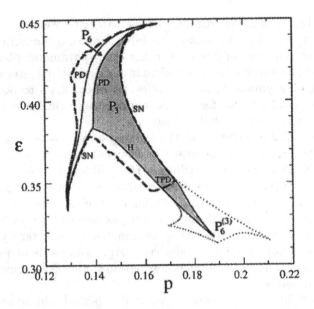

Figure 9. Detailed structure of the period-3 window for $a = 4$. Stability region of the two-cluster period-3 cycle P_3 within the cluster subspace is shaded gray. The boldly dashed curve bounds the region where the cluster attractor $A^{(2)}$ originating from the cycle P_3 is stable in the whole N-dimensional phase space of system (1).

(1). The boldly dashed curve bounds the region of its stability in the whole N-dimensional phase space of system (1). This curve was obtained by a calculation of the transverse Lyapunov exponents (4) for the considered two-cluster states evaluated for a typical trajectory $\{(x(n), y(n))\}_{n=0}^{\infty} \subset A^{(2)}$.

By calculating the transverse Lyapunov exponents (4) we convince ourself that the stability of a two-cluster periodic cycle within the cluster subspace in many cases implies its stability in the whole N-dimensional phase space of system (1). However, the same is not the case for the period-3 cycle considered for $a = 4$ (see Fig. 9). Here, the two-cluster period-3 cycle, while being stable within the cluster subspace (gray region), may be unstable in the whole N-dimensional phase space (gray region outside the boldly dashed curve).

When the parameter point crosses the curve TPD in Fig. 9, the stable asynchronous period-3 cycle $P_3^{(a)}$ undergoes a transverse period-doubling bifurcation, giving rise to a stable period-6 cycle $P_6^{(3)}$ that does not belong to the two-cluster subspace (x, y). This bifurcation occurs when the transverse Lyapunov exponent $\lambda_{\perp,2}^{(2)}$ of the period-3 cycle $P_3^{(a)}$ becomes positive.

As a result, the cluster y is no longer stable but splits into two subclusters. A stable 3-cluster state is born with period-6 temporal in-cluster behavior. In this way, a transverse period-doubling bifurcation can lead to a *cluster-splitting* phenomenon, where the number of synchronized clusters grows by 1. The stability region for the 3-cluster period-6 cycle $P_6^{(3)}$ in the whole N-dimensional phase space of system (1) is bounded by the dotted curve in Fig. 9.

It is worth to note the sensitive dependence of the existence of asymmetric two-cluster states on system size N. For example, for $a = 4$ and $\varepsilon = 0.43$ (see Fig. 9), the two-cluster period-3 cycle $P_3^{(a)}$ gives rise to stable two-cluster states for system (1) if $0.145 \lesssim p \lesssim 0.15$. It follows that system (1) will have stable period-3 two-cluster states, for example, for $N = 100, 101, 102, 103$ ($N_2 = 15$) but not for $N = 99$ or $N = 104$. The large N is the more possible partitions into two-cluster states can be observed. The approach used in the present paper, namely to consider the in-cluster map (3) with continuous asymmetry parameter p and calculate the transverse Lyapunov exponents (4), allows us to avoid direct numerical simulations of the full system (1). For a more detailed consideration of the emergence of asymmetric clusters also for the Hénon map, we refer the reader to Ref. [44].

6. Cluster-splitting bifurcation

Suppose that system (1) has a stable $CkPm(N_1 : N_2 : \ldots : N_K)$-state, i.e, a K-cluster state with period-m temporal dynamics and with the distribution of the elements among the clusters as $N_1 : N_2 : \ldots : N_K$. Then the map F_K of the form (6) with the asymmetry parameters $p_i^{(K)} = N_i/N$, has a stable period-m cycle $\gamma_m^{(K)}$. In-cluster stability of the cycle (i.e., stability with respect to the K-dimensional system F_K) is determined by K in-cluster multipliers $\nu_i^{(K)}$, $i = \overline{1, K}$. The remaining multipliers $\mu_i^{(K)}$, $i = \overline{1, K}$, as for the N-dimensional system (1), calculated in accordance with (7) are transverse. They control stability of $\gamma_m^{(K)}$ with respect to out-of-cluster perturbations. The cycle $\gamma_m^{(K)}$ provides a stable K-cluster states if all its multipliers $\nu_i^{(K)}$ and $\mu_i^{(K)}$, $i = \overline{1, K}$, are less than 1 in absolute value.

When the parameters a and ε of system (1) vary, the multipliers of $\gamma_m^{(K)}$ may leave the unit disk. In this connection, we distinguish between two possibilities: either one of the in-cluster multipliers $\nu_i^{(K)}$ or one of the transverse multipliers $\mu_i^{(K)}$ becomes larger than 1 in absolute value. In the first case, the in-cluster attractor $\gamma_m^{(K)}$ bifurcates within the cluster subspace Π_K changing in-cluster dynamics. In the second case, which is a subject of our interest in this section, the in-cluster cycle $\gamma_m^{(K)}$ losses its stability in the transverse to Π_K directions causing cluster splitting.

Assume that, with varying a parameter, one of the transverse multipliers of the period-m cycle $\gamma_m^{(K)} \in \Pi_K$, namely $\mu_i^{(K)}$ becomes less then -1. The cycle $\gamma_m^{(K)}$ bifurcates transversely to Π_K giving birth to a period-$2m$ cycle $\gamma_{2m}^{(K+1)}$ belonging a cluster subspace $\Pi_{(K+1)} \supset \Pi_K$. If this new-born cycle appears to be stable in the whole N-dimensional phase space of the original system (1), then the transition from K- to $(K+1)$-cluster dynamics occurs, which may be called a *cluster-splitting bifurcation*. In the bifurcation, the ith cluster of N_i elements splits into two new sub-clusters. Let $N_{1,i}$ and $N_{2,i}$ be numbers of the oscillators in each of the new-born sub-clusters. It is clear that $N_{1,i} + N_{2,i} = N_i$. We find that the $(K+1)$-cluster states appeared is stable under the following conditions:

$$\frac{1}{2} \leq \frac{N_{1,i}}{N_{2,i}} \leq 2. \tag{10}$$

Analytical derivation of the formula (10) in the case $C1P1$ clustered state can be found in Ref. [45].

We illustrate different types of the cluster-splitting bifurcation evaluating numerically system (1) of $N = 100$ coupled logistic maps $f(x) = ax(1 - x)$. Choose the nonlinearity parameter $a = 3.84$. Then the map f has an attracting period-3 cycle γ_3. The coherent, one-cluster state given by γ_3 will be stable in the whole N-dimensional phase space if the transverse multipliers $\mu_1^{(1)} = \nu(1 - \varepsilon)^3$ of $\gamma_3 \in D_N$ calculated in accordance with (7) is less than 1 in absolute value. This takes place for a range of the coupling parameter $\varepsilon \in (\varepsilon^-; \varepsilon^+)$, where $\varepsilon^- \approx -0.0454$ and $\varepsilon^+ \approx 2.0454$. Thus, for this range of ε, system (1) has the asymptotically stable coherent state with period-3 temporal dynamics.

With decreasing ε beyond ε^-, the transverse multiplier $\mu_1^{(1)}$ of γ_3 becomes less then -1 and the coherent state losses its stability. After the bifurcation, system (1) demonstrates a variety of two-cluster states of different ratios of cluster sizes $N_1 : N_2$ and with period-6 temporal dynamics. At further decreasing the coupling parameter ε, these clustered states appear to split again in a similar way.

In Fig. 10, we plot examples of one-parameter bifurcation diagrams originated from the stable period-3 coherent state of system (1) following the evolution of the dynamics as ε decreases. During the calculations, for each next value of ε, the initial conditions were slightly randomly perturbed from the asymptotic state of the previous value, and then iterated according to (1) without perturbations.

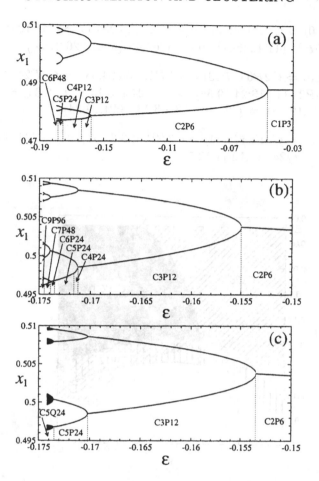

Figure 10. Bifurcation diagrams for the system of coupled logistic maps (1) with $N = 100$ and $a = 3.84$. The only one branch of the original coherent periodic cycle is shown. When calculating, small random perturbations of the amplitude 10^{-9} were applied to the initial conditions at each next value of ε.

Three characteristic examples of cluster-splitting cascades are shown. As we can see, the bifurcation sequences can run in different ways, which depends on previous cluster sizes and on the perturbations noted above. In Fig. 10, $CkPm$ $(N_1 : N_2 : \ldots : N_k)$ denotes a stable k-cluster state with period-m temporal dynamics and with N_j elements in the jth cluster ($CkQm$ states for superposition of a period-m cycle with quasiperiodic dynamics). The following cluster-adding sequences are presented:

(a): $C1P3$ (100) $\Rightarrow$ $C2P6$ (49:51) $\Rightarrow$ $C3P12$ (23:26:51) $\Rightarrow$ $C4P12$ (23:26:21: 30) $\Rightarrow$ $C5P24$ (11:12:26:21:30) $\Rightarrow$ $C6P48$ (5:6:12:26:21:30).

(b): $C1P3$ (100) $\Rightarrow$ $C2P6$ (47:53) $\Rightarrow$ $C3P12$ (23:24:53) $\Rightarrow$ $C4P24$ (10:13:24: 53) $\Rightarrow$ $C5P24$ (10:13:24:19:34) $\Rightarrow$ $C6P24$ (10:13:3:21:19:34) $\Rightarrow$ $C7P48$ (5:5:13:3:21:19:34) $\Rightarrow$ $C9P96$ (2:3:2:3:13: 3:21:19:34).

(c): $C1P3$ (100) $\Rightarrow$ $C2P6$ (46:54) $\Rightarrow$ $C3P12$ (23:23:54) $\Rightarrow$ $C5P24$ (11:12:11: 12:54) $\Rightarrow$ $C5Q24$ (11:12:11:12:54).

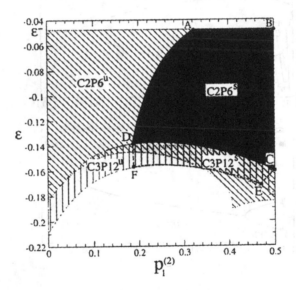

Figure 11. Cluster-splitting bifurcation diagram for the coherent (one-cluster) state of system (1). The dark gray domain $ABCD$ is the stability region of two-cluster states with period-6 temporal dynamics and partition $\{p_1^{(2)}, p_2^{(2)}\}$, where $p_i^{(2)} = N_i/N$, $i = 1, 2$. The light gray domain $DCEF$ is the stability region of three-cluster period-12 states with partition $\{p_1^{(3)}, p_2^{(3)}, p_3^{(3)}\}$, where $p_1^{(3)} = p_2^{(3)} = \frac{1}{2}p_1^{(2)}$ and $p_3^{(3)} = p_2^{(2)}$. Obliquely and vertically hatched regions are for the stability of the two- and three-cluster periodic attractors within the corresponding cluster subspaces, respectively. Parameter $a = 3.84$.

Stability condition (10) is an important characteristics of cluster-splitting bifurcation caused by the period doubling. It is demonstrated in Fig. 11, where the two-parameter diagram of cluster-splitting bifurcation for the coherent (one-cluster) state of system (1) is presented. For $\varepsilon \in (\varepsilon^-; \varepsilon^+)$ (values $\varepsilon^\pm$ are such as above), system (1) has a stable period-3 coherent state. With decreasing ε, this state bifurcates at $\varepsilon = \varepsilon^-$ via transverse supercritical period-doubling bifurcation giving birth to a two-cluster state generated by period-6 cycle $\gamma_6^{(2)}$. The region of the in-cluster stability of $\gamma_6^{(2)}$ is obliquely hatched in Fig. 11. Condition (10) fulfills on the upper segment AB. In-cluster stability region of $\gamma_6^{(2)}$ contains two subregions (with respect to the transverse stability of $\gamma_6^{(2)}$): $C2P6^S$ and $C2P6^U$. In the transverse stability region $C2P6^S$, which is shaded by dark gray, both transverse multipliers $\mu_1^{(2)}$ and $\mu_2^{(2)}$ of the cycle $\gamma_6^{(2)}$ are less than 1 in absolute value. The state is transversally stable forming a stable period-6 two-cluster state of system (1).

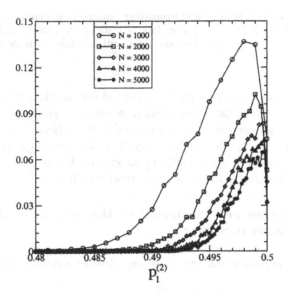

Figure 12.   Probability for a trajectory of system (1) to be attracted by a two-cluster state with partition $\left\{p_1^{(2)}, p_2^{(2)}\right\}$ after transverse destabilization the coherent state. The average has been done over 10000 initial conditions randomly distributed in a small neighborhood (diameter is 10^{-9}) of the unstable coherent state for ε slightly beyond its transverse period-doubling bifurcation. Parameter $a = 3.84$.

At the bifurcation, all values of $p_1^{(2)} \in [1/3, 1/2]$ are possible (see (10) and segment AB in Fig. 11), but their relative probabilities are different, as shown in Fig. 12. As our numerical experiments approve, the most probable clusters to appear in direct calculations with randomly distributed initial conditions are close to symmetric ones.

With further decreasing ε, the next cluster-splitting bifurcation caused by period doubling can occur. Following the cluster-splitting cascades, we summarize our findings in Fig. 13, where all observed cluster transitions

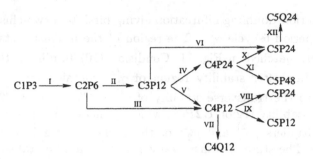

Figure 13. Schematic representation of bifurcation sequences and emergence of clustered states that take place in system (1) after loss of stability by periodic coherent state. $CkPm$ denotes k-cluster states with period-m temporal dynamics (Q is for quasiperiodic motion).

up to five successive cluster splits are collected schematically. One can see that the cluster splitting can proceed with or without temporal period doubling. The later case corresponds to so-called *transcritical cluster splitting*. Number of clusters can also increase by 2 which can happen when the previous cluster splitting was strictly symmetrical. For more details about cluster-splitting bifurcation, we refer the reader to Ref. [45].

7. Synchronization and clustering in the system of diffusively coupled Rössler oscillators

In Sec. 7–9 we consider the chain of the Rössler systems coupled in a diffusive way

$$\dot{u}_j = f(u_j) + C(u_{j+1} + u_{j-1} - 2u_j), \quad j = 1, ..., N \tag{11}$$

and with periodic boundary condition $u_{N+1} = u_1$. Here $u_j \in \mathbb{R}^n$ denote the phase space coordinates of the individual oscillator and C is the coupling matrix. Each of the uncoupled Rössler oscillators

$$\begin{pmatrix} \dot{x}_1 \\ \dot{x}_2 \\ \dot{x}_3 \end{pmatrix} = \dot{X} = f(X) = \begin{pmatrix} -x_2 - x_3 \\ x_1 + ax_2 \\ b + x_3(x_1 - c) \end{pmatrix} \tag{12}$$

is known to have an invariant attracting chaotic set A for the parameter values $a = 0.42$, $b = 2.0$, $c = 4.0$ [47]. Then the synchronization manifold $D = \{u_1 = u_2 = \cdots = u_N\}$ is invariant and contains the invariant chaotic set

$$A_s = \{u_1 = \cdots = u_N \in A\}. \tag{13}$$

Let us recall that the complete synchronization for the time-continuous systems (11) takes place if this "synchronous" set $A_s \subset D$ is asymptotically stable. This implies that small deviations from the state (13) tend to zero, i.e., $\|u_i - u_j\| \to 0$, $\forall i, j$ as $t \to \infty$ for any initial conditions $U(0) = (u_1(0), \ldots, u_N(0))$ from some neighborhood $\mathcal{U} \supset A_s$.

System of the form (11) with coupling only via the first components (i.e., $C = \mathrm{diag}\{\alpha, 0, 0\}$) was considered by Heagy et al. [48]. They discussed an associated size instability which restricts the number of oscillators capable of sustaining stable synchronous chaos even for large coupling. They also developed a general approach, involving the so-called "master stability function" for the investigation of different linear coupling schemes [49, 50]. Phase synchronization effect in a nonidentical array of diffusively coupled Rössler oscillators with a coupling matrix $C = \mathrm{diag}\{0, \alpha, 0\}$ has been investigated by Osipov et al. [51].

By contrast to complete synchronization as defined above, in the case of partial synchronization the coupled system splits into groups of identically oscillating elements called clusters.

In Sec. 8 we obtain conditions for complete synchronization and riddling for the system of three coupled Rössler oscillators. We also prove that this system admits partial synchronization for some narrow parameter range. The system of four coupled oscillators is considered in Sec. 9. It is shown that partial synchronization in such a system can be archived for a "massive" range of the parameters. Conditions to determine when complete or partial synchronization take place are obtained. Finally, we discuss the case of a large number of diffusively coupled oscillators.

First, consider the system of two coupled chaotic oscillators:

$$\begin{aligned}
\dot{u}_1 &= f(u_1) + C(u_2 - u_1), \\
\dot{u}_2 &= f(u_2) + C(u_1 - u_2).
\end{aligned} \tag{14}$$

Synchronization effects for this system are well studied (see, for example the surveys in [52, 53, 54]). For simplicity, we shall consider coupling with only one parameter in the form $C = \alpha \cdot I$ where I is the unit matrix. For the synchronization manifold $u_1 = u_2$, which is invariant for system (14), denote the transverse coordinates by $\xi = u_1 - u_2$.

One-dimensional bifurcation diagram for the desynchronization of low-periodic orbits, riddling and blowout bifurcation is presented in Fig. 14. The variation of the largest transverse Lyapunov exponent of the synchronized chaotic attractor versus α for two coupled Rössler oscillators is depicted as bold curve. The intersection of this graph with the horizontal axis (the point α_2) determines the moment when the attractor becomes unstable in average. The thin lines show the same quantity for individual periodic orbits embedded in the attractor. The rightmost point of intersection of these lines

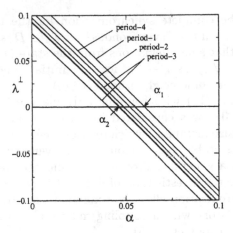

Figure 14.   Largest transverse Lyapunov exponent versus the coupling parameter α, calculated for the chaotic attractor (bold curve) and for some low-periodic orbits.

with the axis (the point α_1) gives an approximation of the moment of the riddling bifurcation. This corresponds to the transverse destabilization of the period-1 orbit. As it follows from the numerical calculations, $\alpha_1 \approx 0.060$ and $\alpha_2 \approx 0.042$.

Figure 15 shows a schematic diagram of the bifurcations associated with transverse destabilization of the period-1 unstable periodic orbit embedded into the attractor A. It illustrates the coexistence of different kinds of synchronous and asynchronous dynamics. For details we refer the reader to [55].

8. Three coupled oscillators

The symmetry properties of the system (11) with three oscillators

$$
\begin{aligned}
\dot{u}_1 &= f(u_1) + \alpha I(-2u_1 + u_2 + u_3), \\
\dot{u}_2 &= f(u_2) + \alpha I(-2u_2 + u_1 + u_3), \\
\dot{u}_3 &= f(u_3) + \alpha I(-2u_3 + u_1 + u_2)
\end{aligned}
\tag{15}
$$

imply that the synchronous set loses its transverse stability in all transverse directions simultaneously. In order to show this, following [49] let us rewrite system (15) in the form

$$
\dot{u} = F(u) + \alpha(G \otimes I)u.
\tag{16}
$$

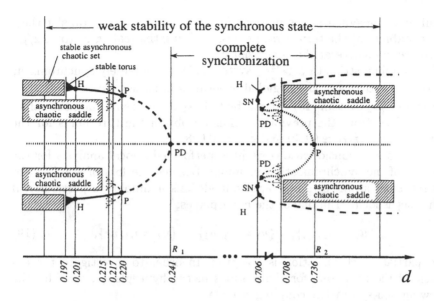

Figure 15.   One-dimensional bifurcation diagram with respect to a coupling parameter. Horisontal line in the center of the diagram corresponds to the period-1 UPO. The bifurcations, which are associated with the UPO are shown. In particular, period-doubling (PD) and pitchfork (P) transverse bifurcations determine riddling moments for the attractor. For the detailed explanations, see [55].

Here, $F(u) = (f(u_1), f(u_2), f(u_3))^T$, $u = (u_1, u_2, u_3)^T$, and the matrix

$$G = \begin{pmatrix} -2 & 1 & 1 \\ 1 & -2 & 1 \\ 1 & 1 & -2 \end{pmatrix}. \tag{17}$$

The variational equation for the synchronized solutions of (16) can be reduced to three systems of the form

$$\dot{\xi} = (Df(s) + \mu_i \alpha I)\xi, \tag{18}$$

where $\mu_i, i = 1, 2, 3$ are the eigenvalues of G, and $s(t)$ is a trajectory on $\mathcal{A}$. In our case $\mu_1 = 0$ and $\mu_2 = \mu_3 = -3$. A detailed explanation of the reduction procedure can be found in [49]. Each of the above equations corresponds to some "transverse mode" (i.e., it determines the behavior of transverse perturbations restricted to some direction) except for that for $\mu_1 = 0$. Hence, the equations for both transverse modes are the same.

Now we can observe that the transverse variational equation for two coupled systems (14) also has the form (18) with $\mu = -2$. Hence, taking into account that $\alpha\mu_i I = -2\alpha \left(\frac{\mu_i}{-2}\right) I$, we can use the results for the local stability of the synchronous motions for two oscillators and transfer them to the stability of the transverse modes, applying the scaling factor $-2/\mu_i$.

Hence, we arrive at the conclusions:

1) For coupling strength $\alpha > \frac{2}{3}\alpha_1$ system (15) is completely synchronized;

2) For $\frac{2}{3}\alpha_2 < \alpha < \frac{2}{3}\alpha_1$ either global or local riddling occurs;

3) For $\alpha < \frac{2}{3}\alpha_2$ system (15) can not be fully synchronized. Using the values of α_1 and α_2 from the previous section we obtain the thresholds for the coupled triplet: $\frac{2}{3}\alpha_1 \simeq 0.040$ and $\frac{2}{3}\alpha_2 \simeq 0.028$.

In the above considerations we have performed a local analysis for the stability of the synchronization manifold. Besides the loss of stability for this set, some stable sets may arise outside this manifold. In particular, if such a set is located in one of the hyperplanes:

$$\{u_1 = u_2, u_3\}, \quad \{u_1 = u_3, u_2\}, \quad \{u_2 = u_3, u_1\}, \tag{19}$$

then partial synchronization is observed. In order to investigate the existence of the limit sets for the motions in the hyperplanes, consider the following non-symmetric coupling scheme

$$\begin{aligned} X' &= f(X) + \alpha(Y - X), \\ Y' &= f(Y) + 2\alpha(X - Y) \end{aligned} \tag{20}$$

which is obtained by the factorization $u_1 = u_2 = X$ and $u_3 = Y$.

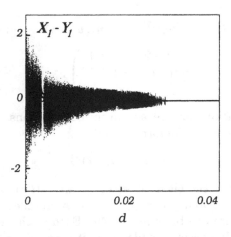

Figure 16.   Bifurcation diagram for the Poincaré map, defined by system (10).

Consider Poincaré return map for system (20) at the point $(0.44, 0, 0, 0, 0, 0)$ with normal vector directed along the X_1-axis. Calculations shows that this map is defined for all parameter α-values in the considered region as well as for the considered initial values. The bifurcation diagram in Fig. 16 shows the evolution of $X_1 - Y_1$ for this map after skipping 300 iterations. We may assume that this procedure reveals the limit sets of our map in the $X_1 - Y_1$-projection and explains the dynamics inside the hyperplanes (19). We clearly observe the loss of synchronization at $\alpha \approx 0.028$ as predicted by the above linear theory. Some periodic windows can also be visible.

Let us examine transverse stability of the limit sets which are located inside the hyperplane. In order to estimate this stability numerically, the variational equation for transverse perturbations is needed. Due to the symmetry it is enough to consider the case $u_1 = u_2 = X, u_3 = Y$. Denote the transverse coordinates $\xi = u_1 - u_2$. Then the variational equation for ξ takes the form

$$\delta\xi' = [Df(X(t)) - 3\alpha]\delta\xi, \tag{21}$$

where $X(t)$ is a solution of (20).

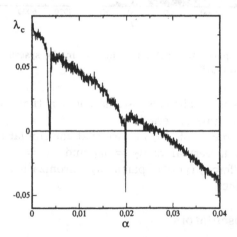

Figure 17.   Maximal Lyapunov exponent that determines the stability of the partially synchronous motion in the system of three coupled Rössler oscillators..

Graph of the maximal Lyapunov exponent λ_c for system (21) is shown in Fig. 17. At the point $\alpha \approx 0.028$ we observe loss of the transverse stability in agreement with the linear theory.

It is interesting to note two narrow intervals around the points $\alpha \approx 0.0038$ and $\alpha \approx 0.0198$, where λ_c becomes negative. These parameter values correspond to the case when limit sets which are located inside the hyperplane $u_1 = u_2$ become stable in the directions transverse to the hyperplane. This implies clustering which happens when the in-cluster dynamics, i.e. the dynamics within the hyperplane $u_1 = u_2$ becomes periodic and asymmetric with respect to $X_1 = Y_1$. Figure 18 shows the evolution of the differences

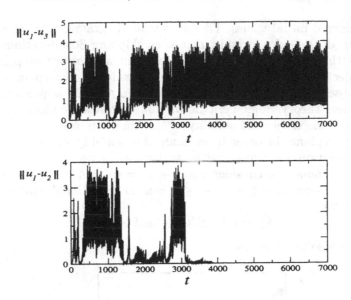

Figure 18. Example of partially synchronous motion in the system of three coupled Rössler oscillators for $\alpha = 0.0198$.

$\|u_1 - u_2\|$ and $\|u_1 - u_3\|$. The calculations confirm that in this case the partially synchronized state is periodic.

Note that the symmetry of (15) implies that the similar stable periodic orbit exists in other hyperplanes: $u_2 = u_3$ and $u_1 = u_3$ for the given parameter values. Different types of partial synchronization can be realized by varying initial values.

9. Four coupled oscillators

In this section we shall show that in the case of four coupled oscillators partial synchronization is observed for more wide parameters ranges. This is related to the fact that the synchronous set (13) first loses its transverse stability in some directions while the other transverse directions remain stable. The system of four coupled oscillators can be written in the form (16) with $u = (u_1, u_2, u_3, u_4)^T$, $F(u) = (f(u_1), \ldots, f(u_4))^T$, and the matrix

$$G = \begin{pmatrix} -2 & 1 & 0 & 1 \\ 1 & -2 & 1 & 0 \\ 0 & 1 & -2 & 1 \\ 1 & 0 & 1 & -2 \end{pmatrix}. \tag{22}$$

The matrix (22) has the eigenvalues $\mu_{1,2} = -2$ and $\mu_3 = -4$ which correspond to different transverse modes. To find these modes we calculate corresponding eigenvectors v_i of the matrix G: $v_1 = (0, 1, 0, -1)$, $v_2 = (1, 0, -1, 0)$, and $v_3 = (-1, 1, -1, 1)$. The first two modes involve the coincidence of $u_2 = u_4$ and $u_1 = u_3$, respectively. Their stability properties are described by equation (18) with $\mu = -2$. The third mode corresponds to the relation $u_1 + u_3 - u_2 - u_4 = 0$ and its stability is determined by equation (18) with $\mu = -4$.

Referring to the coupled Rössler systems, the corresponding quantities will be the following: $\alpha_1 = 0.060$, $\alpha_2 = 0.042$, $\frac{\alpha_1}{2} = 0.030$, and $\frac{\alpha_2}{2} = 0.021$. Figure 19 shows transverse Lyapunov exponents for a typical orbit in the synchronous set. The largest Lyapunov exponent λ_1 corresponds to the longitudinal behavior confined to the synchronization manifold. Therefore this exponent does not depend on α. The next two exponents λ_2 and λ_3 are equal, they correspond to the transverse modes related to the eigenvalues $\mu_{1,2}$ of matrix G. The most stable mode corresponds to the Lyapunov exponent λ_4.

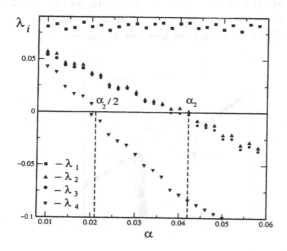

Figure 19.   Scan of four largest Lyapunov exponents versus coupling α for the system of four coupled chaotic Rössler oscillators.

With the stability loss of the first transverse mode, asynchronous attractors arise away from the synchronization manifold. As in the case of three coupled systems, consider the stability of the asynchronous attractor with respect to perturbations which drive the system out of the partially synchronous state. Let examine the following clustering structure

$$X := u_1 = u_2, \quad Y := u_3 = u_4, \tag{23}$$

which comes from the stability analysis of the synchronous set corresponding to the least stable mode.

In a standard way we obtain equations for the "perturbations" from the clustering structure. For this, consider transverse coordinates $\xi_1 = u_1 - u_2$ and $\xi_2 = u_3 - u_4$ which measure the deviations of the trajectory from the clustered manifold (23). The linearized equations admit the form

$$\begin{aligned}
\delta\xi_1' &= [Df(X(t)) - 3\alpha]\delta\xi_1 - \alpha\delta\xi_2 \\
\delta\xi_2' &= [Df(Y(t)) - 3\alpha]\delta\xi_2 - \alpha\delta\xi_1
\end{aligned} \tag{24}$$

where $X(t)$ and $Y(t)$ are solutions of two coupled systems describing the dynamics in the manifold. Largest Lyapunov exponent for the system (24) is responsible for the stability of the linearized system (24), and therefore, for the stability of the partially synchronous motion (23). Figure 20 presents graph of the Lyapunov exponents versus α. As it can be seen, after the loss of complete synchronization at α_2, there exists rather wide parameter range around $\alpha = 0.03$ where λ_c is negative. This corresponds to the existence of a stable partially synchronous structure (23). Due to the symmetry, both clustering structures $u_1 = u_2$, $u_3 = u_4$ and $u_1 = u_4$, $u_2 = u_3$ can be simultaneously realized with varying the initial values.

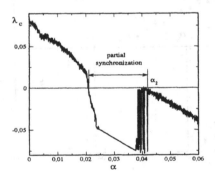

Figure 20.   Maximal Lyapunov exponent that determines stability of the partially synchronous motion $u_1 = u_2$, $u_3 = u_4$ in the system of four coupled Rössler oscillators.

To obtain equations for the in-cluster motions, we substitute $u_1 = u_2 = X$ and $u_3 = u_4 = Y$ (or $u_1 = u_4 = X$ and $u_2 = u_3 = Y$) in Sys. (11) of four oscillators. The resulting equations are exactly of the form (14). Therefore, the in-cluster motions coincide with the motions of the two coupled system (14), see Sec. 7. Hence, the transition to the partial synchronization happens when the system of two coupled Rössler oscillators lost its transverse stability and get a stable asynchronous cycle. We obtain periodic asynchronous motion with $u_1 = u_2$ and $u_3 = u_4$ or with $u_1 = u_4$ and $u_2 = u_3$, depending on the initial conditions. Figure 21 illustrates both possibilities for $\alpha = 0.035$.

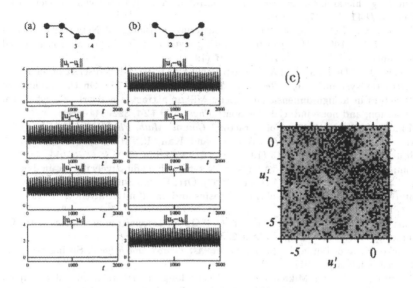

Figure 21. Two asymptotic modes of the partially synchronized behavior for the system of four diffusively coupled Rössler systems. Coupling parameter is $\alpha = 0.035$. Depending on the initial conditions one can get (a) $u_1 = u_2$ and $u_3 = u_4$ or (b) $u_1 = u_4$ and $u_2 = u_3$. (c) Cross-section of the basin of attraction of two stable limit cycles representing different clustering structures.

To clarify the situation, note that actually there coexist two stable symmetric cycles in the phase space of the coupled system. One of them belongs to the manifold $\{u_1 = u_2, u_3 = u_4\}$ and the other to $\{u_1 = u_4, u_2 = u_3\}$. Their basins of attraction. are found to be strongly intermixed. To illustrate this, we have calculated the basins in a two-dimensional cross-section defined as $u_{1i} = 1.0$, $u_{4i} = 1.0$, $u_{2j} = 1.0$, $u_{3j} = 1.0$, where $i = 1, 2, 3, 4$, $j = 2, 3, 4$. The grid was 70×70. The result is shown in Fig. 21c. Small black squares correspond to initial conditions from which the system converges to one of the cycles, leading to the first clustering mode and white points correspond to the second variant of clustering (cf. Fig. 21a,b). For more details we refer the reader to Ref. [56].

References

1. Kaneko, K., "Chaotic but regular posi-nega switch among coded attractors by cluster-size variation," *Phys. Rev. Lett.* **63**, 219-223 (1989); "Clustering, coding, switching, hierarchical ordering, and control in a network of chaotic elemsnts," *Physica D* **41**, 137–172 (1990).
2. Kaneko, K., "Relevance of Dynamics Clustering to Biological Networks," Physica D **75**, 55–73 (1994); "Coupled maps with growth and death: An approach to cell differentiation," *Physica D* **103**, 505–527 (1997).
3. Kaneko, K., "Dominance of Milnor attractors and noise-induced selection in a multiattractor system," *Phys. Rev. Lett.* **78**, 2736–2739 (1997); "On the strength of attractors in a high-dimensional system: Milnor attractor network, robust global attraction, and noise-induced selection," *Physica D* **124**, 322–344 (1998).
4. Milnor, J., "On the concept of attractor," *Comm. Math. Phys.* **99**, 117–195 (1985).
5. Alexander, J.C., Yorke, J.A., You, Z., and Kan, I., "Riddled basins," *Int. J. Bifurcation Chaos* **2**, 795–813 (1992).
6. Sommerer, J.C. and Ott, E., "A physical system with qualitatively uncertain dynamics," Nature (London) **365**, 138–140 (1993); Ott, E. and Sommerer, J.C., "Blowout bifurcations: the occurrence of riddled basins and on-off intermittency," *Phys. Lett. A* **188**, 39–47 (1994).
7. Platt, N., Spiegel, E.A., and Tresser, C., "On-off intermittency: a mechanism for bursting," *Phys. Rev. Lett.* **70**, 279–282 (1993).
8. Kaneko, K. and Tsuda, I., *Complex systems: Chaos and beyond*, Springer-Verlag, Berlin Heidelberg (2001).
9. Manrubia, S.C. and Mikhailov, A.S., "Very long transients in globally coupled maps," *Europhys. Lett.* **50**, 580–586 (2000); "Globally coupled logistic maps as dynamical glasses," *Europhys. Lett.* **53**, 451–457 (2001).
10. Shimada, T. and Kikuchi, K., "Periodicity manifestations in the turbulent regime of the globally coupled map lattice," *Phys. Rev. E* **62**, 3489–3503 (2000).
11. Balmforth, N.J., Jacobson, A., and Provenzale, A., "Synchronized family dynamics in globally coupled maps," *Chaos* **9**, 738–754 (1999).
12. Xie, F. and Hu, G., "Clustering dynamics in globally coupled map lattices," *Phys. Rev. E* **56**, 1567-1570 (1997); Xie, F. and Cerdeira, H.A., "Clustering bifurcation and spatiotemporal Intermittency in RF-driven Josephson junction series arrays," *Int. J. Bifurcation Chaos* **8**, 1713–1718 (1998).
13. Glendinning, P., "The stability boundary of synchronized states in globally coupled dynamical systems," *Phys. Lett. A* **259**, 129–134 (1999); "Transitivity and blowout bifurcations in a class of globally coupled maps," *Phys. Lett. A* **264**, 303–310 (1999).
14. Chate, H. and Losson, J., "Non-trivial collective behavior in coupled map lattices: A transfer operator perspective," *Physica D* **103**, 51–72 (1997).
15. Shibata, T. and Kaneko, K., "Collective chaos," *Phys. Rev. Lett.* **81**, 4116–4119 (1998).
16. Cencini, M., Falcioni, M., Vergni, D., and Vulpiani, D., "Macroscopic chaos in globally coupled maps," *Physica D* **130**, 58–72 (1999).
17. Grassberger, P., "Synchronization of coupled systems with spatiotemporal chaos," *Phys. Rev. E* **59**, R2520–R2522 (1999).
18. Topaj, D., Kye, W.-H., and Pikovsky, A., "Transition to coherence in populations of coupled chaotic oscillators: a linear response approach," *Phys. Rev. Lett.* **87**, 074101 (2001).

19. Pikovsky, A.S. and Kurths, J., "Collective behavior in ensembles of globally coupled maps," *Physica D* **76**, 411–419 (1994).

20. Hamm, A., "Large deviations from the thermodynamic limit in globally coupled maps," *Physica D* **142**, 41–69 (2000).

21. Pikovsky, A., Rosenblum, M., and Kurths, J., "Synchronization in a population of globally coupled chaotic oscillators," *Europhys. Lett.* **34**, 165–170 (1996).

22. Zanette, D. and Mikhailov, A.S., "Condensation in globally coupled populations of chaotic dynamical systems," *Phys. Rev. E* **57**, 276–281 (1998).

23. Gelover-Santiago, A.L., Lima, R., and Martinez-Mekler, G., "Synchronization and cluster periodic solutions in globally coupled maps," *Phys. A* **283**, 131–135 (2000).

24. Chatterjee, N. and Gupte, N., "Analysis of spatiotemporally periodic behavior in lattices of coupled piecewise monotonic maps," *Phys. Rev. E* **63**, 017202 (2001).

25. Chaté, H. and Manneville, P., "Collective behaviors in coupled map lattices with local and nonlocal connections," *Chaos* **2**, 307–313 (1992).

26. Ouchi, N.B. and Kaneko, K., "Coupled maps with local and global interactions," *Chaos* **10**, 359–365 (2000).

27. Belykh, V., Belykh, I., Komrakov, N., and Mosekilde, E., "Invariant manifolds and cluster synchronization in a family of locally coupled map lattices," *Discr. Dyn. Nat. Soc.* **4**, 245–256 (2000).

28. Shuai, J.W., Wong, K.W., and Cheng L.M., "Synchronization of spatiotemporal chaos with positive conditional Lyapunov exponents," *Phys. Rev. E* **56**, 2272-2275 (1997).

29. Zhou, C. and Lai, C.-H., "Analysis of spurious synchronization with positive conditional Lyapunov exponents in computer simulations," *Physica D* **135**, 1-23 (2000).

30. Pikovsky, A., Popovych, O., and Maistrenko, Yu., "Resolving clusters in chaotic ensembles of globally coupled identical oscillators," *Phys. Rev. Lett.* **87**, 044102 (2001).

31. Wang, W., Kiss, I.Z., and Hudson, J.L., "Experiments on arrays of globally coupled chaotic electrochemical oscillators: synchronization and clustering," *Chaos* **10**, 248–256 (2000); "Clustering of arrays of chaotic chemical oscillators by feedback and forcing," *Phys. Rev. Lett.* **86**, 4954–4957 (2001).

32. Miyakawa, K. and Yamada, K., "Synchronization and clustering in globally coupled salt-water oscillators," *Physica D* **151**, 217–227 (2001).

33. Vanag, V.K., Yang, L., Dolnik, M., Zhabotinsky, A.M., and Epstein, I.R., "Oscillatory cluster patterns in a homogeneous chemical systems with global feedback," *Nature (London)* **406**, 389–391 (2000).

34. Pollmann, M., Bertram, M., and Hinrich Rotermund, H., "Influence of time delayed global feedback on pattern formation in oscillatory CO oxidation on Pt(110)," *Chem. Phys. Lett.* **346**, 123–128 (2001).

35. Maistrenko, Yu.L., Maistrenko, V.L., Popovych, O., and Mosekilde, E., "Transverse instability and riddled basins in a system of two coupled logistic maps," *Phys. Rev. E* **57**, 2713-2724 (1998).

36. Maistrenko, Yu.L., Maistrenko, V.L., Popovych, O., and Mosekilde, E., "Unfolding of the riddling bifurcation," *Phys. Lett. A* **262**, 355–360 (1999); "Desynchronization of chaos in coupled logistic maps," *Phys. Rev. E* **60**, 2817–2830 (1999).

37. Ashwin, P., Buescu, J., and Stewart, I., "Bubbling of attractors and synchronisation of chaotic oscillators," *Phys. Lett. A* **93**, 126–139 (1994); "From attractor to chaotic saddle: a tale of transverse instability," *Nonlinearity* **9**, 703–737 (1996).

38. Lai, Y.-C., Grebogi, C., Yorke, J.A., and Venkataramani, S.C., "Riddling bifurcation in chaotic dynamical systems," *Phys. Rev. Lett.* **77**, 55–58 (1996).

39. Mira, C., Gardini, L., Barugola, A., and Cathala, J.-C., *Chaotic Dynamics in Two-Dimensional Noninvertible Maps* World Scientific, Singapore, (1996).

40. Maistrenko, Yu.L., Maistrenko, V.L., Popovych, O., and Mosekilde, E.,"Role of the absorbing area in chaotic synchronization," *Phys. Rev. Lett.* **80**, 1638-1641 (1998).

41. Popovych, O., Maistrenko, Yu., Moseskilde,E., Pikovsky, A., and Kurths, J., "Transcritical loss of synchronization in coupled chaotic systems," *Phys. Lett. A* **275**, 401–406 (2000); "Transcritical riddling in a system of coupled maps," *Phys. Rev. E* **63**, 036201 (2001).

42. Hasler, M., Maistrenko, Yu., and Popovych, O., "Simple example of partial synchronization of chaotic systems," *Phys. Rev. E* **58**, 6843-6846 (1998); Maistrenko, Yu., Popovych, O., and Hasler, M., "On strong and weak chaotic partial synchronization," *Int. J. Bifurcation Chaos* **10**, 179–203 (2000).

43. Popovych, O., Maistrenko, Yu., and Mosekilde, E., "Loss of coherence in a system of globally coupled maps," *Phys. Rev. E* **64**, 026205 (2001).

44. Popovych, O., Maistrenko, Yu., and Mosekilde, E., "Role of asymmetric clusters in desynchronization of coherent motion," *Phys. Lett. A* **302**, 171–181 (2002).

45. Popovych, O., Pikovsky, A., and Maistrenko, Yu., "Cluster-splitting bifurcation in a system of coupled maps," *Physica D* **168**, 106–125 (2002).

46. Kaneko, K., "Information cascade with marginal stability in a network of chaotic elements," *Physica D* **77**, 456–472 (1994).

47. Rössler, O.E., "Equation for continuous chaos," *Phys. Lett. A* **57**, 397–398 (1976).

48. Heagy, J., Pecora, L., and Carroll, T., "Short wavelength bifurcation and size instabilities in coupled oscillator systems," *Phys. Rev. Lett.* **74**, 4185–4188 (1995).

49. Pecora, L. and Carroll, T., "Master stability functions for synchronized coupled Systems," *Phys. Rev. Lett.* **80**, 2109–2112 (1998).

50. Pecora, L., "Synchronization conditions and desynchronization patterns in coupled limit-cycle and chaotic systems," *Phys. Rev. E* **58**, 347–360 (1998).

51. Osipov, G.V., Pikovsky, A.S., Rosenblum, M.G., and Kurths, J., "Phase synchronization effects in a lattice of nonidentical Rössler oscillators," *Phys. Rev. E* **55**, 2353–2361 (1997).

52. Rul'kov, N.F.,"Images of synchronized chaos: Experiments with circuits," *Chaos* **6**, 262–279 (1996).

53. Pecora, L.M., Carroll, T.L., Johnson, G.A., Mar, D.J., and Heagy, J.F., "Fundamentals of synchronization in chaotic systems, concepts, and applications," *Chaos* **7**, 520-543 (1997).

54. Yanchuk, S., Maistrenko, Yu., and Mosekilde, E., "Synchronization of time-continuous chaotic oscillators," *Chaos* **13**, (2003).

55. Yanchuk, S., Maistrenko, Yu., and Mosekilde, E., "Loss of synchronization in coupled Rössler systems," *Physica D* **154**, 26–42 (2001).

56. Yanchuk, S., Maistrenko, Yu., and Mosekilde, E., "Partial synchronization and clustering in a system of diffusively coupled chaotic oscillators," *Math. Comp.Simul.* **54**, 491–508 (2001).

NONLINEAR PHENOMENA IN NEPHRON-NEPHRON INTERACTION

E. MOSEKILDE[‡], O. V. SOSNOVTSEVA[††] and N.-H.HOLSTEIN-RATHLOU[*]

[‡]*Department of Physics, The Technical University of Denmark, 2800 Kgs. Lyngby, Denmark*
[†]*Department of Physics, Saratov State University, Astrakhanskaya str. 83. Saratov, 410026, Russia*
[*]*Department of Medical Physiology, University of Copenhagen, 2200 Copenhagen N, Denmark*

1. Introduction

By controling the excretion of water and salts, the kidneys play an important role in regulating the blood pressure and maintaining a proper environment for the cells of the body. This control depends to a large extent on mechanisms that are associated with the individual functional unit, the nephron. However, a variety of cooperative phenomena arising through interactions among the nephrons may also be important. The purpose of this chapter is to present experimental evidence for a coupling between nephrons that are connected via a common piece of afferent arteriole, to develop a mathematical model that can account for the observed synchronization phenomena, and to discuss the possible physiological significance of these phenomena. We are particularly interested in synchronization effects that can occur among neighboring nephrons that individually display irregular (or chaotic) dynamics in their pressure and flow regulation.

It has long been recognized that the nephrons can compensate for variations in the arterial blood pressure. This ability rests partly with the so-called tubuloglomerular feedback (TGF) mechanism by which the individual nephron can regulate the incoming blood flow in dependence of the ionic composition of the fluid leaving the loop of Henle [1]. Early experiments by Leyssac and Baumbach [2] and by Leyssac and Holstein-Rathlou [3, 4] demonstrated that this feedback regulation can become unstable and

A. Pikovsky and Y. Maistrenko (eds.), Synchronization: Theory and Application, 139–174.

generate self-sustained oscillations in the proximal intratubular pressure with a typical period of 30-40 s. With different amplitudes and phases the same oscillations can be observed in the distal intratubular pressure and in the chloride concentration near the terminal part of the loop of Henle [5]. While for normal rats the oscillations have the appearance of a limit cycle with a sharply peaked power spectrum reflecting the period of the cycle, highly irregular oscillations, displaying a broadband spectral distribution with significant subharmonic components, were observed for spontaneously hypertensive rats (SHR) [3].

It has subsequently been demonstrated [6, 7] that irregular oscillations can occur for normal rats as well, provided that the arterial blood pressure is increased by ligating the blood flow to the other kidney (so-called 2 kidney-1 clip Goldblatt hypertension). In a particular experiment, where the function of the nephron was accidentally disturbed, evidence of a period-doubling transition was observed [8]. Together with the above mentioned subharmonic components in the spectral distribution for the hypertensive rats, this type of qualitative change in behavior supports the view that the pressure and flow regulation in the rat nephron operates close to a transition to deterministic chaos [9, 10].

As illustrated in the schematic drawing of Fig. 1, the TGF regulation is made possible by the interesting anatomical feature that the terminal part of the ascending limb of the loop of Henle passes within cellular distances of the afferent arteriole for the same nephron. At the point of contact, specialized cells (the macula densa cells) monitor the NaCl concentration of the tubular fluid and produce a signal that activates the smooth muscle cells in the arteriolar wall. The higher the glomerular filtration is, the faster the fluid will flow through the loop of Henle, and the higher the NaCl concentration will be at the macula densa. A high NaCl concentration causes the macula densa cells to activate the vascular smooth muscle cells in the arteriolar wall and thus to reduce the diameter of that vessel. Hence, the blood flow and, thereby, the glomerular filtration are lowered. The net result is that the TGF mechanism acts as a negative feedback control.

The steady state response of the arteriolar flow regulation can be obtained from open-loop experiments [11] in which a paraffin block is inserted into the middle of the proximal tubule and the rate of filtration is measured as a function of an externally forced flow of artificial tubular fluid into the loop of Henle. Reflecting physiological constraints, this response follows an S-shaped characteristics with a maximum at low Henle flows and a lower saturation level at externally forced flows beyond $20 - 25$ nl/min. The steepness of the response is significantly higher for spontaneously hypertensive rats than for normotensive rats [12]. Together with the delay in the TGF regulation, this steepness plays an essential role for the stability

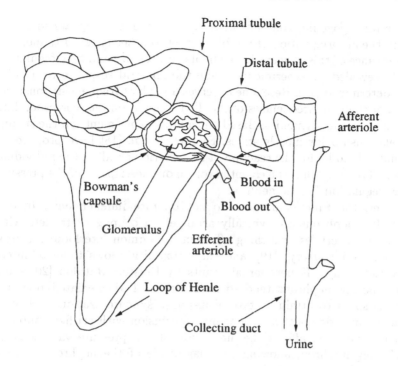

Figure 1.   Sketch of the main structural components of the nephron. Note particularly how the terminal part of the loop of Henle passes within cellular distances of the afferent arteriole, allowing the TGF mechanism to control the incoming blood flow in response to the ionic composition of the fluid leaving the loop of Henle.

of the feedback system [5, 13]. The experimentally observed higher steepness for spontaneously hypertensive rats may therefore explain the more complicated pressure variations observed in these rats.

A main component in the regulatory delay is associated with the finite transit time of the fluid through the tubular system. The length of this delay can be estimated from the phase shift between the pressure oscillations in the proximal tubule and the oscillations of the NaCl concentration in the distal tubule. A typical value is 10-15 *s* [14]. In addition there is a transmission time of 3-5 *s* for the signal from the macula densa cells to reach the smooth muscle cells in the arteriolar wall [5, 14]. In total this delay is sufficient for the nephrons in normotensive rats to operate close to or slightly beyond a Hopf bifurcation [13, 15]. Hollenberg and Sandor [16] have provided indirect evidence that similar oscillations occur in man.

Besides reacting to the TGF signal, the afferent arteriole also responds to variations in the pressure difference across the arteriolar wall. This response consists of a passive elastic component in parallel with an active muscular (or myogenic) component. A similar response appears to be involved in the autoregulation of the blood flow to other types of tissue, and the significance of this element in the nephron pressure and flow regulation is clearly revealed in experiments where the spectral response to a noise input is determined [17]. Here, one observes a peak at frequencies considerably higher than the frequencies of the TGF regulation and corresponding to typical arteriolar dynamics. Based on *in vitro* experiments on the strain-stress relationship for muscle strips, Feldberg *et al.* [18] have proposed a mathematical model for the reaction of the arteriolar wall in the individual nephron. This model plays an essential role in our description of the pressure and flow regulation for the nephron.

However, the functional units do not operate independently of one another. The nephrons are typically arranged in couples or triplets with their afferent arterioles branching off from a common interlobular artery (or cortical radial artery) [19], and this proximity allows them to interact in various ways. Experimental results by Holstein-Rathlou [20] show how neighboring nephrons tend to adjust their TGF-mediated pressure oscillations so as to attain a state of in-phase synchronization. Holstein-Rathlou has also demonstrated how microperfusion with artificial tubular fluid in one nephron affects the amplitude of the pressure variations in a neighboring nephron, allowing the magnitude of the nephron-nephron interaction to be estimated [20].

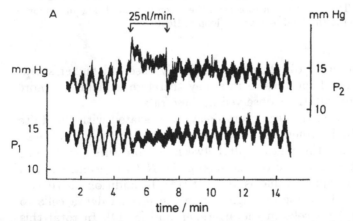

Figure 2. Results of a microperfusion experiment for a pair of neighboring nephrons. Arrows indicate the start and stop of the perfusion phase. In the microperfused nephron (top trace), the tubular pressure oscillations are blocked during the perfusion. During the same period, the amplitude of the oscillations are reduced in the nonperfused nephron (lower trace).

As an illustration of these results, Fig. 2 shows how microperfusion into the proximal tubule of one nephron can influence the pressure oscillations in a neighboring nephron. In the microperfused nephron (top trace) the proximal tubular pressure oscillations are blocked during the microperfusion. Arrows indicate the start and stop of the perfusion phase. During the same period, the amplitude of the oscillations is decreased in the non-perfused nephron (lower trace). Note how the oscillations are reactivated simultaneously in both nephrons, and how in-phase synchronization between the nephrons is reestablished after a few minutes. This type of cross-talk among the nephrons is assumed to be produced by signals that are transmitted along the afferent arterioles [20]. The mechanisms underlying such a coupling are not known in detail. However, two different types of interaction are likely:

(i) A communication between the TGF mechanisms of neighboring nephrons. The presence of such an interaction is well-established experimentally, but the underlying cellular mechanisms remain less well understood. The coupling is associated with a vascular propagated response where electrical signals, initiated by the TGF of one nephron, travel across the smooth muscle cells in the arteriolar wall from the region close to the macula densa and upstream along the arteriole to the branching point with the arteriole from the neighboring nephron. Because of the relatively high speed at which such signals propagate as compared with the length of the vessels and the period of the TGF-mediated oscillations, this type of coupling tends to produce in-phase synchronization. If the afferent arteriole of one nephron is stimulated by the TGF-mechanism to contract, the vascular signals almost immediately reach the neighboring nephron and cause it to contract as well.

(ii) A much simpler type of coupling that we shall refer to as hemodynamic coupling. This coupling arises from the fact that if one nephron is stimulated by its TGF-mechanism to contract its afferent arteriole, then the hydrostatic pressure rises over the neighboring nephron, and the blood flow to this nephron increases. Half a period later when the increased blood flow activates the TGF-mechanism of the neighboring nephron and causes its afferent arteriole to contract, the blood flow to this nephron is again reduced, and the blood flow to the first nephron increases. This type of coupling tends to produce out-of-phase or anti-phase synchronization between the pressure oscillations of the two nephrons.

In reality, we expect both mechanisms to be present simultaneously. Depending on the precise structure of the arteriolar network this may cause one mechanism to be the more important for the local coupling of closely situated nephrons, while the other mechanism might be important for more long range coupling phenomena. Let us also note that simulation results for systems of interacting nephrons were published already by Jensen et al. [8] and by Bohr et al. [21]. These studies describe a variety of different synchronization patterns including a chess-board pattern of anti-phase synchronization for nephrons arranged in a square lattice. However, at that time the physiological mechanisms underlying the nephron-nephron interaction were not yet understood. The present discussion of interacting nephrons is based on our recent publications [22, 23, 24, 25, 26]. A survey of some of the results on synchronization of the pressure and flow control between neighboring nephrons may be found in a recent book on Chaotic Synchronization [27].

2. Single-Nephron Model

Our model of the individual nephron [10, 22] considers the proximal tubule as an elastic structure with little or no flow resistance. The pressure P_t in the proximal tubule changes in response to differences between the in- and outgoing fluid flows

$$\frac{dP_t}{dt} = \frac{1}{C_{tub}} [F_{filt} - F_{reab} - F_{Hen}]. \tag{1}$$

Here F_{filt} is the glomerular filtration rate, F_{reab} represents the reabsorption that takes place in the proximal tubule, F_{Hen} is the flow of fluid into the loop of Henle, and C_{tub} is the elastic compliance of the tubule. The Henle flow,

$$F_{Hen} = \frac{P_t - P_d}{R_{Hen}}, \tag{2}$$

is determined by the difference between the proximal (P_t) and the distal (P_d) tubular pressures and by the flow resistance R_{Hen}. This description is clearly a simplification, since a significant reabsorption of water and salts occurs as the fluid passes through the loop of Henle. However, within the physiologically relevant flow range (2) provides a good approximation to the experimentally determined pressure-flow relation [9].

As the filtrate flows into the descending limb of the loop of Henle, the NaCl concentration in the fluid surrounding the tubule increases significantly, and osmotic processes cause water to be reabsorbed. At the same time, salts and metabolic biproducts are secreted into the tubular fluid. In the ascending limb, on the other hand, the tubular wall is nearly impermeable to water. Here, the epithelial cells contain molecular pumps that transport sodium and chloride from the tubular fluid into the space between the nephrons (the interstitium). These processes are accounted for in considerable detail in the spatially extended model developed by Holstein-Rathlou et al. [13]. In the present model, the reabsorption F_{reab} in the proximal tubule and the flow resistance R_{Hen} are treated as constants.

The glomerular filtration rate is given by [28]

$$F_{filt} = (1 - H_a) \left(1 - \frac{C_a}{C_e}\right) \frac{P_a - P_g}{R_a},$$
(3)

where the afferent hematocrit H_a represents the fraction that the blood cells constitute of the total blood volume at the entrance to the glomerular capillaries. C_a and C_e are the protein concentrations of the afferent and efferent blood plasma, respectively, and R_a is the flow resistance of the afferent arteriole. $(P_a - P_g)/R_a$ determines the incoming blood flow. Multiplied by $(1 - H_a)$ this gives the plasma flow. Finally, the factor $(1 - C_a/C_e)$ expresses conservation of proteins in the blood during passage of the glomerular capillaries.

The glomerular pressure P_g is determined by distributing the arterial to venous pressure drop between the afferent and the efferent arteriolar resistances, and the protein concentration C_e in the efferent blood is obtained from the assumption that filtration equilibrium is established before the blood leaves the glomerular capillaries. This leads to an expression of the form

$$C_e = \frac{1}{2b} \left[\sqrt{a^2 - 4b(P_t - P_g)} - a \right],$$
(4)

where a and b are parameters relating the colloid osmotic pressure to the protein concentration [29].

The glomerular feedback is described by a sigmoidal relation between the muscular activation ψ of the afferent arteriole and the delayed Henle flow

$$\psi = \psi_{max} - \frac{\psi_{max} - \psi_{min}}{1 + \exp\left[\alpha\left(3x_3/TF_{Hen0} - S\right)\right]}.$$
(5)

This expression is based on empirical results for the variation of the glomerular filtration with the flow into the loop of Henle as obtained in the above mentioned open-loop experiments [12]. In Eq. 5, $\psi_{max} - \psi_{min}$ denotes the dynamical range of the muscular activity as controled by the TGF mechanism. α determines the slope of the feedback curve. We have already indicated that this slope plays an important role for the stability of the pressure and flow regulation. In the next section we shall use α as one of the main bifurcation parameters. S is the displacement of the curve along the normalized flow axis, and F_{Hen0} is a normalization value for the Henle flow.

The delay in the tubuloglomerular feedback is taken into account by means of a chain of three first-order coupled differential equations,

$$\frac{dx_1}{dt} = F_{Hen} - \frac{3}{T}x_1,$$
(6)

$$\frac{dx_2}{dt} = \frac{3}{T}(x_1 - x_2),\tag{7}$$

$$\frac{dx_3}{dt} = \frac{3}{T}(x_2 - x_3),\tag{8}$$

with T being the total delay time. In this way the delay is represented as a smoothed process, with x_1 and x_2 being intermediate variables in the delay chain and with $3x_3/T$ being the delayed value of F_{Hen}.

The afferent arteriole is divided into two serially coupled sections of which the first (representing a fraction β of the total length) is assumed to have a constant hemodynamic (i.e., flow) resistance, while the second (closer to the glomerulus) is capable of varying its diameter and hence the flow resistance in dependence of the tubuloglomerular feedback activation,

$$R_a = R_{a0}\left[\beta + (1 - \beta)\,r^{-4}\right].\tag{9}$$

Here, R_{a0} denotes a normal value of the arteriolar resistance and r is the radius of the active part of the vessel, normalized relatively to its resting value. The hemodynamic resistance of the active part is assumed to vary inversely proportional to r^4. This is an application of Poiseuille's law for laminar flows.

Experiments have shown that arterioles tend to perform damped, oscillatory contractions in response to external stimuli [30]. This behavior may be captured by the set of two coupled first-order differential equations

$$\frac{dv_r}{dt} + kv_r - \frac{P_{av} - P_{eq}}{\omega} = 0, \quad \frac{dr}{dt} = v_r\tag{10}$$

Here, k is a characteristic rate constant describing the damping of the arteriolar dynamics, and ω is a parameter that controls the natural frequency of the oscillations. P_{av} is the average pressure in the active part of the arteriole, and P_{eq} is the value of this pressure for which the arteriole is in equilibrium with its present radius at the existing muscular activation.

As previously noted, the reaction of the arteriolar wall to changes in the blood pressure consists of a passive, elastic component in parallel with an active, muscular response. The elastic component is determined by the properties of the connective tissue [31]. The active component in the strain-stress relation appears to be surprisingly simple. For some value ϵ_{max} of the strain ϵ, the active stress attains a maximum, and on both sides the stress decreases almost linearly with $|\epsilon - \epsilon_{max}|$. Moreover, the stress is proportional to the muscle tone ψ. By numerically integrating the passive and active contributions across the arteriolar wall, one can establish a relation among the equilibrium pressure P_{eq}, the normalized radius r, and the activation level ψ [18]. Unfortunately, computation of this relation for every time step of the simulation model is quite time consuming. To speed up the process we have used an approximation in the form of the expression [22]

$$P_{eq} = 2.4 \times e^{10(r-1.4)} + 1.6(r-1) + \psi \left(\frac{4.7}{1 + e^{13(0.4-r)}} + 7.2(r+0.9) \right), \quad (11)$$

where P_{eq} is expressed in kPa ($1 \; kPa = 10^3 \; N/m^2 \cong 7.5 \; mmHg$). The first two terms in (11) represent the pressure *vs.* radius relation for the nonactivated arteriole. The terms proportional to ψ represent the active response. This is approximately given by a sigmoidal term superimposed onto a linear term. The activation from the TGF mechanism is assumed to be determined by (5). The expression in (11) closely reproduces the prediction of the more complex, experimentally based relation [22].

The above equations complete our description of the single-nephron model. A more detailed account of the physiological processes underlying the model may be found in Topics in Nonlinear Dynamics [10] or in the paper by Barfred *et al.* [22]. In total we have six coupled ordinary differential equations, each representing an essential physiological relation. Because of the need to numerically evaluate C_e in each integration step, the model cannot be brought onto an explicit form. The parameters applied in the single-nephron model may also be found in our previous publications [10, 22].

Figure 3 shows an example of a one-dimensional bifurcation diagram for the single-nephron model obtained by varying the slope α of the open-loop response characteristics (5) while keeping the other parameters constant. In particular, the delay in the feedback regulation is assumed to be $T = 16 \; s$. The diagram was constructed by combining a so-called brute force bifurcation diagram with a diagram obtained by means of continuation methods [32, 33]. Such methods allow us to follow stable as well as unstable periodic orbits for a nonlinear dynamical system under variation of a parameter and to identify the various bifurcations that the orbits undergo. Hence, in Fig. 3 fully drawn curves represent stable solutions and dotted curves represent unstable periodic solutions.

For a given value of α, the brute force bifurcation diagram displays all the values of the relative arteriolar radius r that the model attains when the steady state trajectory intersects a given Poincaré section in phase space. To reveal the coexistence of several stable solutions, the brute force diagram has been obtained by scanning α in both directions.

For $T = 16 \; s$, the single nephron model undergoes a supercritical Hopf bifurcation at $\alpha \cong 11$ (outside the figure). In this bifurcation, the equilibrium point loses its stability, and stable periodic oscillations smoothly emerge as the steady-state solution. For $\alpha \cong 19.5$, at the point denoted PD_a^{1-2} in Fig. 3, the simple periodic oscillations undergo a period-doubling

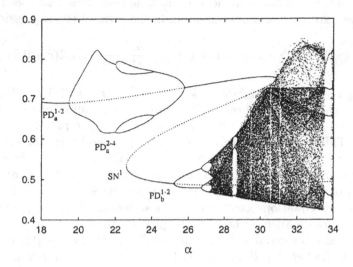

Figure 3.   One-dimensional bifurcation diagram for the single-nephron model obtained by varying the slope of the open-loop response characteristics. r represents the normalized arteriolar radius. $T = 16$ s. Dotted curves represent unstable solutions determined by means of continuation techniques. Two saddle-node bifurcations of the period-1 cycle fold an incomplete period-doubling structure over a complete period-doubling transition to chaos.

bifurcation, and in a certain interval of α-values the period-2 cycle is the only stable solution. As we continue to increase α, the period-2 solution undergoes a new period-doubling bifurcation at $\alpha \cong 22$ (i.e., at the point denoted PD_a^{2-4}). The presence of a stable period-4 cycle is revealed in Fig. 3 by the fact that r assumes four different values for the same value of α.

With further increase of α, the stable period-4 orbit undergoes two consecutive backwards period-doublings, so that the original period-1 cycle again becomes stable around $\alpha = 26$. The stable period-1 cycle exists up to $\alpha \cong 31$ where it is destabilized in a saddle-node bifurcation. The saddle cycle can be followed backwards in the bifurcation diagram (dotted curve) to a point near $\alpha = 22.5$ where it undergoes a second saddle-node bifurcation, and a new stable period-1 orbit is born. This cycle has a considerably larger amplitude than the original period-1 cycle. As the parameter α is again increased, the new period-1 cycle undergoes a period-doubling cascade starting with the first period-doubling bifurcation at $\alpha \cong 25$ and accumulating with the development of deterministic chaos near $\alpha = 27$. At even higher values of α we notice, for instance, the presence of a period-

3 window near $\alpha = 28.5$, and the appearance of a stable period-4 cycle around $\alpha = 33.5$.

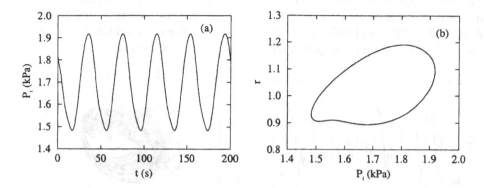

Figure 4. (a) Temporal variation of the proximal tubular pressure P_t as obtained from the single-nephron model for $\alpha = 12$ and $T = 16$ s. (b) Corresponding phase plot. With the assumed parameters the model displays self-sustained oscillations in good agreement with the behavior observed for normotensive rats. The tubular pressure is given in kPa ($1\ kPa = 7.5\ mmHg$).

The above scenario is typical of nonlinear dynamical systems when the amplitude of the internally generated oscillations becomes sufficiently large. In the bifurcation diagram of Fig. 3 this occurs when the slope of the feedback characteristics exceeds a critical value. However, similar scenarios can be produced through variation of other parameters such as, for instance, the feedback delay. On a general level, the bifurcation diagram also agrees with the experimental observation of a period-doubling in the response of a nephron to an external disturbance.

For normotensive rats, the typical operation point around $\alpha = 10 - 12$ and $T \cong 16$ s falls near the Hopf bifurcation point. This agrees with the finding that in a typical experiment about 70% of the nephrons perform self-sustained oscillations while the remaining show stable equilibrium behavior [5]. We can also imagine how the system is shifted back and forth across the Hopf bifurcation by variations in the arterial pressure. This explains the characteristic temporal behavior of the nephrons with periods of self-sustained oscillations interrupted by periods of stable equilibrium dynamics.

Figure 4(a) shows the variation of the proximal tubular pressure P_t with time as calculated from the single-nephron model for $\alpha = 12$ and $T = 16$ s. With these parameters the system operates slightly beyond the Hopf bifurcation point, and the depicted pressure variations represent the steady-state oscillations reached after the initial transient has died out.

With physiologically realistic parameter values the model thus reproduces the observed self-sustained oscillations with appropriate periods and amplitudes. Figure 4(b) shows the phase plot. Here, we have displayed the normalized arteriolar radius r against the proximal intratubular pressure. Again, the 50% amplitude in the variations of r seems reasonable. Along the limit cycle the motion proceeds in the clockwise direction.

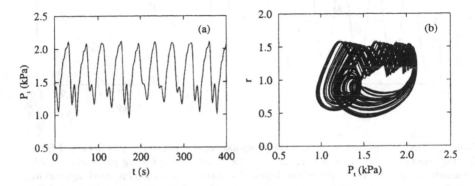

Figure 5.   (a) Pressure variations obtained from the single-nephron model for $\alpha = 32$ and $T = 16\ s$. (b) Corresponding phase plot. With these parameters the model displays chaotic oscillations resembling the behavior observed for spontaneously hypertensive rats [9].

As previously noted, spontaneously hypertensive rats (SHR) have significantly larger α-values than normal rats ($\alpha = 16.8 \pm 12.0$ *vs.* $\alpha = 11.4 \pm 2.2$ for normotensive rats) [12]. On the other hand, it appears that the feedback delay is approximately the same for the two strains. Figure 5(a) shows an example of the chaotic pressure variations obtained for higher values of the TGF response. Here, $\alpha = 32$ and $T = 16\ s$. Under these conditions, the oscillations never repeat themselves and, as calculations show, the largest Lyapunov exponent is positive [34]. The corresponding phase plot in Fig. 5(b) displays the characteristic picture of a chaotic attractor. One can also analyse the behavior in terms of an interplay between the rapid modulations associated with the arteriolar dynamics and the slower TGF-mediated oscillations. The two modes never get into step with one-another, however. We shall return to a discussion of this type of mode interaction and its significance for the synchronization phenomena in Sec. 3.

3. Nephron as a bi-modal oscillator

3.1. WAVELET ANALYSIS

Signals generated by living systems are seldom stationary and homogeneous, and processing of such signals by means of conventional techniques such as Fourier analysis often leads to problems with respect to the interpretation of the obtained results. Among the various approaches developed to study nonstationary data, wavelet analysis is probably the most popular [46]. In particular, this method gives us the possibility to investigate the temporal evolution of signals with different rhythmic components.

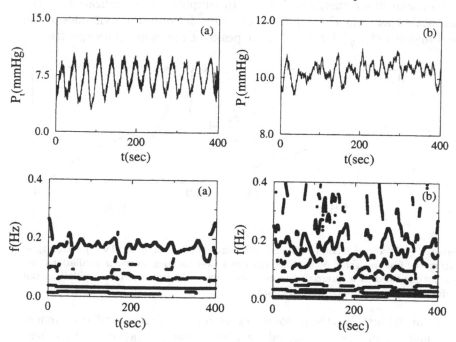

Figure 6.   Upper panel: Regular tubular pressure oscillations from a normotensive rat (a) and irregular pressure variations from a spontaneously hypertensive rat (b). Lower panel: Wavelet analyses showing the main components detected in the two time-series.

The wavelet transform of a signal $x(t)$ is performed as follows:

$$T_\psi[x](a,b) = \frac{1}{\sqrt{a}} \int_{-\infty}^{\infty} x(t)\psi^* \left(\frac{t-b}{a} \right) dt, \qquad (12)$$

where ψ is a "mother" function that in general has a soliton-like shape with zero mean. $T_\psi[x](a,b)$ are the wavelet coefficients. a and b are parameters that measure, respectively, the temporal extension of the wavelet and the timing of its center. In the analysis of signals with various rhythmic contributions, the so-called *Morlet*-wavelet is often applied:

$$\psi(t) = \exp(jk_0 t) \exp\left[-\frac{t^2}{2}\right]. \tag{13}$$

In the present analysis we shall use this function with $k_0 = 2\pi$.

Besides the wavelet transform coefficients $T_\psi[x](a,b)$ we can estimate the energy density $E_\psi[x](a,b) = |\ T_\psi[x](a,b)\ |^2$. As a result there is a 3-dimensional surface $E_\psi[x](a,b)$ or $E_\psi[x](f,b)$, where f is the frequency ($f = 1/a$). Sections of this surface at fixed time moments $b = t_0$ correspond to the momentary energy spectrum. To simplify the visualization of the time-scale spectrum $E_\psi[x](f,b)$ we can consider only the dynamics of the local maxima of $E_\psi[x](f, t_0)$, i.e., the peaks of the momentary spectra.

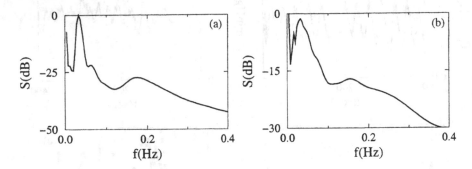

Figure 7.   Power spectrum obtained from the wavelet analysis. Two peaks, representing the fast myogenic oscillations and the slower tubuloglomerular oscillations, are well-distinguished. It is interesting to note how clearly the fast arteriolar oscillations can be discerned in the tubular pressure data.

Figure 6 illustrates the periodicities related to different oscillatory modes and their harmonics as observed for a normotensive (a) and a hypertensive (b) rat, respectively. Inspection of the figure reveals that the slow oscillations, whether they are periodic or chaotic, maintain a nearly constant frequency of about 30 mHz throughout the observation time. The fast oscillations, on the other hand, fluctuate around a significantly higher value (around $180 mHz$). The lack of persistence may be related to the complex modulation of the fast oscillations by the slow dynamics or to the influence of noise. However, Fig. 6 does not give information about the relative amplitudes of the various spectral components. Such information can be obtained from a *scalogram* (i.e., the time averaged power spectrum), being the analogue of the conventional Fourier power spectrum. This is illustrated in Fig. 7 where two well-pronounced peaks around $0.03Hz$ and

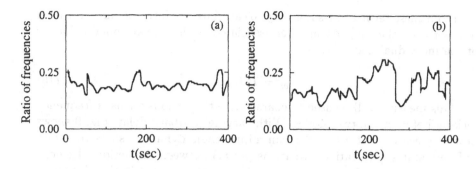

Figure 8.   Relation between the internal time scales for a normotensive rat (a) and for a hypertensive rat (b). The normotensive rat is seen to maintain a near 1:5 synchronization between its fast and slow oscillations throughout the observation period.

$0.15 - 0.2Hz$, respectively, are distinguished. Since the above frequency ranges are of physiological interest we extract these oscillations from the original wavelet transformation for further analysis of their coherent properties. Figure 8 displays the relation between the fast and slow oscillations in a single nephron. For the periodic oscillations observed in normotensive rats, the fast and slow components adjust their periods in accordance with one another to maintain a $1 : 5$ or $1 : 6$ entrainment during the observation time (Fig. 8a). For the chaotic oscillations observed in hypertensive rats, the ratio changes more randomly in time (Fig. 8b).

3.2. SIMULATION

Both the experimental investigations and our simulations reveal one of the most important features of the single-nephron operation, namely the presence of two different time scales in the pressure and flow variations. Considering the model equations (3) we can identify the two time scales in terms of (i) a low-frequency (TGF-mediated) oscillation with a period $T_h \cong 2.2T$ arising from the delay in the tubuloglomerular feedback, and (ii) significantly faster oscillations with a period $T_v \approx T_h/5$ associated with the inherent myogenic adjustment of the arteriolar radius.

To determine T_h and T_v in our numerical simulations we have used the mean return times of the trajectory to appropriately chosen Poincaré sections

$$T_v = \left\langle T_{ret} \big|_{v_r=0} \right\rangle \quad , \quad T_h = \left\langle T_{ret} \big|_{x_2=0} \right\rangle, \tag{14}$$

From these return times it is easy to calculate the intra-nephron rotation number (i.e., the rotation number associated with the two-mode behavior of the individual nephron)

$$r_{vh} = T_v/T_h. \tag{15}$$

This measure will be used to characterize the various forms of frequency locking between the two modes. With varying feedback delay, Fig. 9 shows how the two oscillatory modes can adjust their dynamics so as to attain different states with rational relations $(n : m)$ between the periods. Regions of high resonances (1 : 4, 1 : 5, and 1 : 6) are seen to exist in the physiologically interesting range of the delay time $T \in [12\ s, 20\ s]$. For lower values of α (corresponding to the conditions for normotensive rats) one also find regions with 1:1 and 1:2 synchronization for $T = 16\ s$.

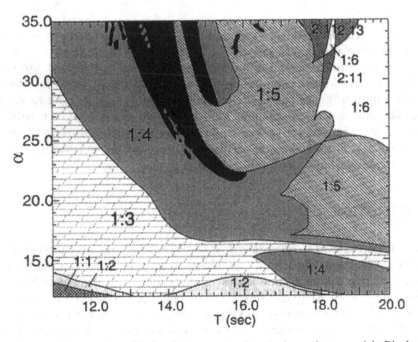

Figure 9.   Two-mode oscillatory behavior in the single nephron model. Black colored regions represent chaotic solutions. For high values of α we find 1:4, 1:5, and 1:6 synchronization.

While the transitions between the different locking regimes always involve bifurcations, bifurcations may also occur within the individual regime. A period-doubling transition, for instance, does not necessarily change r_{vh}, and the intra-nephron rotation number may remain constant through a

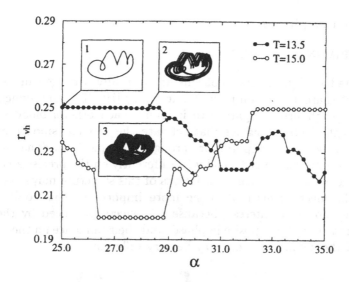

Figure 10. Internal rotation number for a single nephron as a function of the parameter α. It is interesting to note that the internal synchronization can be maintained through a complete period-doubling cascade and into the chaotic regime.

complete period-doubling cascade and into the chaotic regime. This is illustrated in Fig. 10 where we have plotted r_{vh} as a function of the feedback gain α for different time delays $T_2 = 13.5$ s (black circles) and $T = 15.0$ s (open circles). Phase projections (P_t, r) from the various regimes are shown as inserts. Inspection of the figure clearly shows that r_{vh} remains constant under the transition from regular 1:4 oscillations (black circles for α = 25.0) to chaos (for α = 28.0), see inserts 1 and 2. With further evolution of the chaotic attractor (insert 3), the 1:4 mode locking is destroyed. In the interval around α = 31.5 we observe 2:9 mode locking. A similar transition is observed for the 1:5 locked regime (open circles). Periodic oscillations (α = 27.0) evolve into a chaotic attractor (α = 28.5) but the rotation number maintains a constant value. For fully developed chaos (insert 3), the 1:5 locking breaks down.

We conclude that besides being regular or chaotic, the self-sustained pressure variations in the individual nephron can be classified as being synchronous or asynchronous with respect to the ratio between the two time scales that characterize the fast (vascular) mode and the slow (TGF mediated) mode, respectively. As we shall see, this complexity in behavior may play an essential role in the synchronization between a pair of interacting nephrons.

4. Coupled Nephrons

4.1. EXPERIMENTAL RESULTS

As illustrated in Fig. 11, the nephrons are typically arranged in pairs or triplets that share a common interlobular artery [19]. This anatomical feature allows neighboring nephrons to influence each other's blood supply either through electrical signals that activate the vascular smooth muscle cells of the neighboring nephron or through a simple hemodynamic coupling. The two mechanisms depend very differently on the precise structure of the arteriolar network. Hence, variations of this structure may determine which of the mechanisms that is the more important. This could be of considerable biological interest, because the effects produced by the two mechanisms tend to be opposite in phase, and their influence on the overall behavior of the nephron system may be very different.

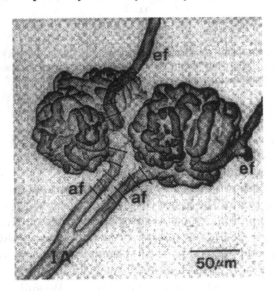

Figure 11. Scanning electron microscope picture of the arteriolar system for a couple of adjacent nephrons. The nephrons are assumed to interact with one another via muscular contractions that propagate along the afferent arterioles (af) and via the so-called hemodynamic coupling.

In order to study the interaction between the nephrons, experiments were performed with normotensive as well as with spontaneously hypertensive rats at the Department of Medical Physiology, University of Copenhagen, and the Department of Physiology, Brown University [37]. During the experiments the rats were anaesthetized, placed on a heated operating table to maintain the body temperature, and connected to a small animal respirator to ensure a proper oxygen supply to the blood. The frequency of the respirator was close to 1 Hz. This component is clearly visible in the frequency spectra of the observed tubular pressure variations. Also observable is the frequency of the freely beating heart, which typically gives a contribution in the 4-6 Hz regime. The frequencies involved in the nephron pressure and flow regulation are significantly lower and, presumably, not influenced much by the respiratory and cardiac forcing signals [9].

When exposing the surface of a kidney, small glass pipettes, allowing simultaneous pressure measurements, could be inserted into the proximal tubuli of a pair of adjacent nephrons. For superficial nephrons the tubules are directly observable on the surface of the kidney when using a microscope. After the experiment, a vascular casting technique was applied to determine if the considered nephron pair shared a common piece of afferent arteriole. Only nephrons for which such a shared arteriolar segment was found showed clear evidence of synchronization, supporting the hypothesis that the nephron-nephron interaction is mediated by the network of incoming blood vessels [19, 23, 38].

In Fig. 12 the left panel shows an example of different types of synchronization. Branching from different arterioles the nephrons in the two top traces operate nearly 180° out of phase. We consider anti-phase synchronization to be the signature of a strong hemodynamic component in the coupling, i.e., contraction of the afferent arteriole for one nephron causes the blood flow to the adjacent nephron to increase. In line with this interpretation, inspection of the arteriolar tree has shown that the nephrons in this case, while sharing an interlobular artery, are too far apart for the vascularly propagated coupling to be active. It is clearly seen, however, that the nephrons in the lower two traces, branching from the same arteriole, demonstrate in-phase synchronization.

Inspection of the right panel of Fig. 12 reveals that the three nephrons (all from the same arteriole) are synchronized and remain nearly in phase for the entire observation period (corresponding to ten periods of oscillation).

In order to investigate the problem of phase synchronization for the complicated pressure variations observed in the system of interacting nephrons we have applied the method introduced by Rosenblum et al. [39, 40]. With this approach one can follow the temporal variation of the difference $\Delta\Phi(t) = \Phi_2(t) - \Phi_1(t)$ between the instantaneous phases $\Phi_1(t)$ and $\Phi_2(t)$ for a pair of coupled chaotic oscillators. The instantaneous phase $\Phi(t)$ and amplitude $A(t)$ for a signal $s(t)$ with complicated (chaotic) dynamics may be defined from

$$A(t)e^{j\Phi(t)} \equiv s(t) + j \,\tilde{s}\,(t) \tag{16}$$

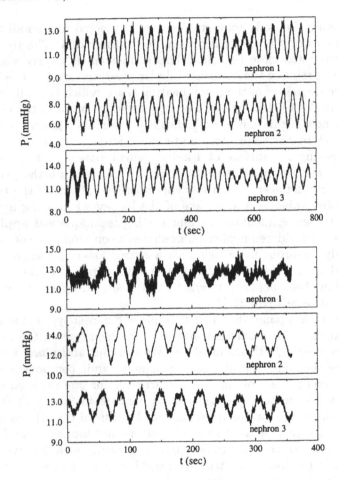

Figure 12. Simultaneous tubular pressure variations for a triple of coupled nephrons in two normotensive rats. The upper panel demonstrates in-phase and anti-phase synchronization in the pressure variations between nephrons that belong to the same, respectively to different, arterioles.

where

$$\tilde{s}\,(t) = \frac{1}{\pi} PV \int\limits_{-\infty}^{\infty} \frac{s(\tau)}{t - \tau} d\tau \qquad (17)$$

denotes the Hilbert transform of $s(t)$, j being the imaginary unit. The notation PV implies that the integral should be evaluated in the sense of Cauchy principal value.

$m : n$ phase synchronization between two oscillators is said to occur if

$$|n\Phi_2(t) - m\Phi_1(t) - C| < \mu \qquad (18)$$

where μ is a small parameter ($\mu < 2\pi$) that controls the allowed play in
the phase locking. In particular, 1:1 phase synchronization is realized if
the phase difference $\Phi_2(t) - \Phi_1(t)$ remains bound to a small interval μ
around a mean value C. For systems subjected to external disturbances
or noise one can only expect the condition for phase synchronization to
be satisfied for finite periods of time, interrupted by characteristic jumps
in $\Delta\Phi$. Under these circumstances one can speak about a certain degree
of phase synchronization if the periods of phase locking are significant
compared to the characteristic periods of the interacting oscillators [41].
Alternatively, one can use the concept of frequency synchronization if the
weaker condition

$$\Delta\Omega = \left\langle n\,\dot{\Phi}_2\,(t) - m\,\dot{\Phi}_1\,(t) \right\rangle = 0 \qquad (19)$$

is satisfied. Here, $\langle\rangle$ denotes time average, and $\Delta\Omega$ is the difference in
(mean) angular frequencies. As noted above, 1:1 frequency synchronization
is already distinguishable from the spectral distribution of the experimental
data.

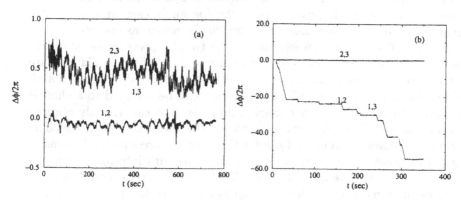

Figure 13.   Relative phase dynamics for pairs of nephrons in normotansive rats. (a)
illustrates in phase ($\Delta\pi \approx 0$) and antiphase synchronization ($\Delta\pi \approx 0.5$) for nephrons with
dominating vascularly propagated and hemodynamic coupling, respectively. (b) shows the
variation of the phase difference for nephrons branching off from the same arteriole.

Figure 13(a) shows variations of the (normalized) phase difference for
the regular pressure variations in a normotensive rat. One clearly observes
in-phase ($\Delta\phi \cong 0$) and antiphase ($\Delta\phi \cong \pi$) synchronization for nephrons
branching from the same and from different arterioles, respectively. All

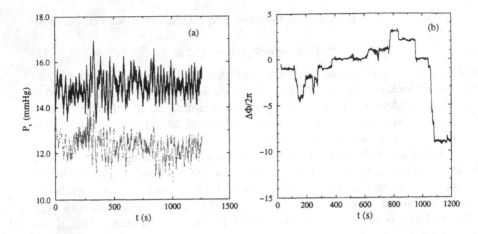

Figure 14.   (a) Example of the tubular pressure variations that one can observe in adjacent nephrons for hypertensive rats and (b) variation of the normalized phase difference $\Delta\phi/2\pi$ for the irregular pressure variations. This is an illustration of the phenomenon of chaotic phase synchronization [27].

three nephrons in the triple (b) demonstrate synchronous behavior but there are characteristic phase slips of 2π (or an integer number of 2π) because of the noisy conditions under which the nephrons operate.

Figures 14 represents an example of the tubular pressure variations in pairs of neighboring nephrons for hypertensive rats. These oscillations are significantly more irregular than the oscillations displayed in Figs. 12 and, as previously discussed, it is likely that they can be ascribed to a chaotic dynamics. In spite of this irregularity, however, one can visually observe a certain degree of synchronization between the interacting nephrons. The normalized phase difference $\Delta\Phi/2\pi$ for the irregular pressure oscillations in Fig. 14(b) reveals characteristic locking intervals with intermediate phase slips. Here, we note the interval from $t \cong 400$ s to $t \cong 800$ s (corresponding to 16 oscillations of the individual nephrons) where the phase difference remains nearly constant.

We conclude that the experimental results show clear evidence of synchronization of neighboring nephrons both for normotensive and for hypertensive rats. Moreover, one can observe both in-phase and anti-phase synchronization, associated, presumably, with two different coupling mechanisms between the nephrons. In the next section we shall study the transitions to and between different regimes of synchronization in the two-nephron model. Particularly interesting in this connection is the role of multistability in the chaotic phase synchronization [42, 43].

4.2. NUMERICAL SIMULATION

The description of synchronization phenomena observed for interacting oscillators may be divided into two stages. The first step is to consider the case when the coupling strength is sufficiently weak so that analytical methods can be applied. The second step is to examine the case of finite coupling strength and show to what extent the results of the weak-coupling limit can be extrapolated. Since the definition of phase multistability involves the phase difference between the interacting oscillators, the phase variables will be the main quantities used to characterize the collective dynamics.

First, let us consider the weak-coupling case, i.e., we assume that the coupling causes only small perturbations of the limit cycles of the uncoupled oscillators. The coupled system may then be approximated by a phase model [47] where the phase ϕ of a limit cycle oscillator is defined by $d\phi(V_0)/dt = 1$ with $V_0 \in R^N$ being a point on the limit cycle. Applying the concept of *isochrons* defined stroboscopically as a subset of initial conditions that asymptotically converge to the same point on the limit cycle [47], the phase description can be extended to some vicinity of the limit cycle. Moreover, for a sufficiently small vicinity of the limit cycle one can assume that the above subset is a flat surface that is transversal to the limit cycle at a given point.

In the presence of a small perturbation $P(V)$, the phase dynamics obeys the following equation [47]:

$$\frac{d\phi}{dt} = 1 + Z(\phi)P(V), \tag{20}$$

where the sensitivity function $Z(\phi) = grad_V\phi|_{V=V_0}$ measures the change of phase along the limit cycle caused by the change of V. Namely, we choose a point V_0 on the limit cycle and a point V close to V_0 but not on the limit cycle and then measure the difference in phases between V_0 and V. In the limit $|V - V_0| \to 0$, this difference, divided by $|V - V_0|$, gives the sensitivity function $Z(\phi)$.

The interaction of two *identical* oscillators with phases ϕ_1 and ϕ_2 can be quantified through the evolution of their phase difference $\Delta\phi = \phi_1 - \phi_2$. In the limit of weak interaction, averaged over a period, the phase dynamics for one of the oscillators can be expressed as [47]

$$\frac{d(\Delta\phi)}{dt} = \Gamma(\Delta\phi) = \frac{1}{2\pi}\int_0^{2\pi} d\phi Z(\phi)P(\phi, \Delta\phi), \tag{21}$$

where $P(\phi, \Delta\phi) = P(V_0(\phi), V_0(\phi + \Delta\phi))$ describes the rate of change of the state vector V of one oscillator due to the interaction with another oscillator with a phase difference $\Delta\phi$, and ZP is the phase shift along the limit cycle for the given perturbation. Note, that the limit cycles in both systems are assumed to have similar shapes, i.e., to be topologically conjugated.

For mutually coupled oscillators, the entrainment manifests itself as a mutual phase shift. This can be analyzed purely in terms of the *antisymmetric* part $\Gamma_a(\Delta\phi)$ of the effective coupling function (21) [47]. The zeroes of $\Gamma_a(\Delta\phi)$ correspond to the phase-locked synchronous states ($\Delta\phi = Const.$) and their stabilities are determined from the slope of $\Gamma_a(\Delta\phi)$ at the respective states, i.e., a negative slope means a stable state, and vice versa. This method of effective coupling has been used in a number of recent investigations [48, 49, 50].

When the coupling becomes strong enough to modify the geometry of the limit cycle, the phase reduction method can no longer be used. Direct numerical methods should then be applied. First of all, we calculate a set of points on the limit cycle modified by the interaction. Over a set of initial conditions covering the full length of the limit cycle, we follow the evolution of the initial phase shift $\Delta\phi(t)$ to some fixed value $\Delta\phi(t+\tau)$. Plotting these results together, i.e. $\Delta\phi(t+\tau)$ *vs.* $\Delta\phi(t)$, we obtain a one-dimensional phase map with a discrete time step τ. Analysis of this map allows us to find the fixed points and estimate their stabilities.

Note that for the effective coupling method one can obtain the phase map in terms of Γ_a. Namely, for two coupled identical oscillators the dynamics of the phase difference is given by [47]:

$$\frac{d(\Delta\phi)}{dt} = 2\Gamma_a(\Delta\phi). \tag{22}$$

Setting $dt \to \tau$ and $d(\Delta\phi) \to (\Delta\phi_{t+\tau} - \Delta\phi_t)$ for small enough τ one finds the expression:

$$\Delta\phi_{t+\tau} \approx \Delta\phi_t + 2\tau\Gamma_a(\Delta\phi_t), \tag{23}$$

to which our numerical calculations converge for vanishing coupling.

Suppose that two models of the form (1-11) are coupled via a vector diffusive coupling where linear terms proportional to the difference between the variables of the two subsystems are introduced in each of the six equations of motion.

Let us consider how a coupling acting through the different variables influences the dynamics of the system. Figure 15 illustrates the antisymmetric parts Γ_a of the effective coupling function calculated separately for the six cases of one-variable coupling. For the variables P_t, x_1, x_2, and x_3 the plots of Γ_a almost coincide and possess a zero point with negative slope at $\Delta\phi = 0$. This means that a single in-phase synchronous regime can be found when two nephron models are diffusively coupled through these variables. It is also clearly seen that the synchronous antiphase regime is unstable ($\Delta\phi = \pi$). Generally, the observed behavior is very natural and expected for a wide class of classical oscillators (like Van der Pol oscillators). Phase multistability can not be detected. However, a coupling via the r or v_r variables reveals a more complex behavior. The Γ_a curve for r has three stable and three unstable equilibrium points. Note that the in-phase regime $\Delta\phi = 0$ is unstable. Similar behavior is observed for v_r, but in this case the in-phase regime is related to a neutral equilibrium point, $d(\Gamma_a)/d(\Delta\phi) \to 0|_{\Delta\phi \to 0}$.

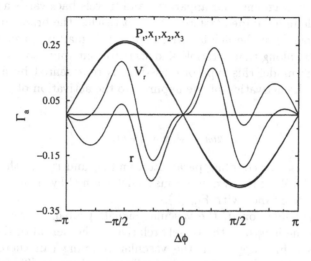

Figure 15. Antisymmetric part of the effective coupling function for the nephron model at $T = 16.0$ s and $\alpha = 18.6$. The behavior of $\Gamma_a(\Delta\phi)$ reveals the synchronization properties of two coupled 6-dimensional oscillators.

Thus, application of the phase reduction method to the relatively high-dimensional nephron model allows us to diagnose a number of qualitatively different responses of this oscillator to varying coupling mechanisms. This is in good agreement with both experimental data and simulation results. It is known that the P_t, x_1, x_2, x_3, and v_r variables are involved in the tubuloglomerular feedback loop that is responsible for low-frequency oscillations with a period $T_h = 2.2T$ (As before, T denotes the delay in the TGF regulation). On the other hand, the r and v_r variables, demonstrating somewhat faster oscillations with a period $T_v \approx T_h/4$, are associated with the adjustment of the arteriolar radius. Hence, different nephron responses to distinct types of coupling are related to the above biological processes.

Let us make a further interesting observation. With a 1:4 ratio between the frequencies of the oscillatory modes, one can expect four coexisting stable regimes for the coupling via r or v_r. However, only three such regimes (zero points with negative slope) can be found in Fig. 15. We believe that this result is due the fact that the oscillatory modes in the single nephron are not actually weakly coupled. Thus, the properties of the point $\Delta\phi = 0$ are defined through the competition of oscillatory modes that can compensate each other even when the coupling takes place via one variable. To get an additional stable point, coupling via the P_t variable should be introduced. This is in agreement with our biological understanding of the model.

The situation is the same for biologically motivated couplings. Let us start by considering the vascular coupling. The muscular activation ψ arises in the so-called juxtaglomerular apparatus and travels backwards along the afferent arteriole in a damped fashion. When it reaches the branching point with the arteriole from the neighboring nephron, it may propagate in the forward direction along that arteriole and start to contribute to its vascular response. In our model this type of cross-talk is represented by adding a contribution of the activation of one nephron to the activation of the other, i.e.,

$$\psi_{1,2tot} = \psi_{1,2} + \gamma\psi_{2,1} \tag{24}$$

where γ is the vascular coupling parameter, and ψ_1 and ψ_2 are the uncoupled activation levels of the two nephrons as determined by their respective Henle flows in accordance with Eq. (5).

As previously mentioned, the vascular signals propagate very fast as compared with the length of the vessels relative to the period of the TGF-oscillations. As a first approach, the vascular coupling can therefore be considered as instantaneous. Experimentally one observes [35] that the magnitude of the activation decreases exponentially as the signal travels along a vessel. Hence, only a fraction of the activation from one nephron can contribute to the activation of the neighboring nephron, and $\gamma = e^{-l/l_0} < 1$. Here, l is the propagation length for the coupling signal, and l_0 is the characteristic length scale of the exponential decay. As a base case value, we shall use $\gamma = 0.2$.

To implement the hemodynamic coupling, a piece of common interlobular artery is included in the system, and the total length of the incoming blood vessel is hereafter divided into a fraction $\varepsilon < \beta$ that is common to the two interacting nephrons, a fraction $1 - \beta$ that is affected by the TGF signal, and a remaining fraction $\beta - \varepsilon$ for which the flow resistance is considered to remain constant. As compared with the equilibrium resistance of the separate arterioles, the piece of shared artery is assumed to have half the flow resistance per unit length.

Defining P_ε as the pressure at the branching point of the two arterioles, the equation of continuity for the blood flow reads

$$\frac{P_a - P_\varepsilon}{\varepsilon R_{a0}/2} = \frac{P_\varepsilon - P_{g1}}{R_{a1}} + \frac{P_\varepsilon - P_{g2}}{R_{a2}} \tag{25}$$

with

$$R_{a1} = (\beta - \varepsilon) R_{a0} + (1 - \beta) R_{a0} r_1^{-4} \tag{26}$$

and

$$R_{a2} = (\beta - \varepsilon) R_{a0} + (1 - \beta) R_{a0} r_2^{-4}. \tag{27}$$

Here, R_{a0} denotes the total flow resistance for each of the two arterioles in equilibrium. r_1 and r_2 are the normalized radii of the active part of the afferent arterioles for nephron 1 and nephron 2, respectively, and P_{g1} and P_{g2} are the corresponding glomerular pressures. As a base case value of the hemodynamic coupling parameter we shall use $\varepsilon = 0.2$.

Because of the implicit manner in which the glomerular pressure is related to the efferent colloid osmotic pressure and the filtration rate, direct solution of the set of seven coupled algebraic equations for the two-nephron system becomes rather inefficient. Hence, for each nephron we have introduced the glomerular pressure P_g as a new state variable determined by

$$\frac{dP_{g,i}}{dt} = \frac{1}{C_{glo}} \left(\frac{P_\varepsilon - P_{g,i}}{R_{a,i}} - \frac{P_{g,i} - P_v}{R_e} - F_{filt,i} \right) \tag{28}$$

with $i = 1, 2$. This implies that we consider the glomerulus as an elastic structure with a compliance C_{glo} and with a pressure variation determined by the imbalance between the incoming blood flow, the outgoing blood flow, and the glomerular filtration rate.

From a physiological point of view, this formulation is well justified. Compared with the compliance of the proximal tubule, C_{glo} is likely to be quite small, so that the model becomes numerically stiff. In the limit $C_{glo} \to 0$, the set of differential equations reduces to the formulation with algebraic equations presented in Sec. 2. Finite values of C_{glo} will change the damping of the system, and therefore also the details of the bifurcation structure. In practice, however, the model will not be affected significantly as long as the time constant $C_{glo} R_{eff}$ is small compared with the periods of interest. Here, R_{eff} denotes the effective flow resistance faced by C_{glo}.

Figure 16 shows a segment of the bifurcation diagram for synchronous solutions on the mismatch vs. hemodynamic coupling parameter plane. The strength of the vasculary propagated interaction is fixed at 0.004. In-phase oscillations are stable when both interacting systems are nearly identical ($T_1 \approx T_2$) and the hemodynamic coupling is weak enough ($\varepsilon < 0.0115$). However, due to the self-modulated nature of the oscillations in the individual nephron, there are also two stable out-of-phase synchronous regimes (O_1 and O_2). When ε increases, the antiphase regime A also becomes stable due to the effect of the hemodynamic coupling. Within some interval of ε, there are four stable coexisting solutions: the in-phase solution I, the antiphase solution A, and two out-of-phase regimes O_1 and O_2. inspection of the figure clearly shows that the synchronization region has a complicated inherent structure. With increasing mismatch, the O_1 and O_2 cycles lose their stability via a tangent bifurcation (entering the nonsynchronous region) or via a period doubling at the border of the PD zone in Fig. 16.

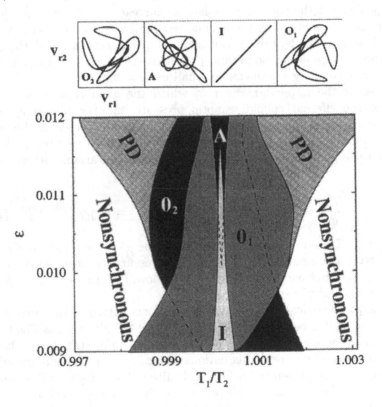

Figure 16. Synchronization regions for coexisting families of attractors ($\alpha = 18.595$, $T_2 = 16.0$ s, $\gamma = 0.004$). In-phase ($\Delta\phi = 0.0$) and anti-phase ($\Delta\phi \approx -1.3474$) solutions are labeled **I** and **A**, respectively. Two out-of-phase regimes with $\Delta\phi \approx -0.7773$ and $\Delta\phi \approx 2.9129$ are indicated as O_1 and O_2. PD denotes regions of period-doubled solutions. The inserts show characteristic phase space projections of the four synchronized solutions.

4.3. ENTRAINMENT OF DIFFERENT OSCILLATORY MODES.

By virtue of the two-mode dynamics of the individual nephron, a number of new and interesting results appear. Using anatomical criteria, neighboring nephrons having a high likelihood of deriving their afferent arteriole from the same interlobular artery were identified [51]. In these nephrons 29 out of 33 pairs (i.e. 80 %) were found to have synchronized oscillations. In contrast, nephron pairs not fulfilling these criteria only showed synchronous oscillations in one case out of 23 investigated pairs (i.e. 4 %). This observation shows that synchronized oscillations are preferentially found in nephrons originating from the same interlobular artery. From time series one can visually observe a certain degree of synchronization between the interacting nephrons. However, it is difficult to separately estimate the degree of adjustment for the myogenic oscillations and for the TGF mediated oscillations without special tools.

To study multimode interactive dynamics in coupled systems we propose to use the wavelet based coherence measure Γ_Δ (in analogy with the classical coherence function). Let $E_\psi[xx](f, t)$ and $E_\psi[yy](f, t)$ be the energy densities for signals $x(t)$ and $y(t)$. Let also in some range of frequencies Δ each of the processes $x(t)$ and $y(t)$ have a clearly expressed rhythm (e.g., range of slow or fast oscillations for nephrons). In this case synchronization means that the corresponding frequencies for $x(t)$ and $y(t)$ will be locked (coincide). Such a situation corresponds to the value $\Gamma_\Delta = 1$ for the function:

$$\Gamma_\Delta^2(t) = \frac{\max_{f \in \Delta} \left[E_\psi[xy](f, t) \right]^2}{\max_{f \in \Delta} \left[E_\psi[xx](f, t) \right] \cdot \max_{f \in \Delta} \left[E_\psi[yy](f, t) \right]}. \tag{29}$$

Here, $E_\psi[xy](f, t)$ is the mutual energy density $E_\psi[xy](f, t) = \mid T_\psi[xy](f, t) \cdot T_\psi^*[yx](f, t) \mid$. $\Gamma_\Delta(t)$ is a function of time that allows us to follow the evolution of the interactive dynamics of the two processes in the chosen frequency range Δ. The more synchronous the rhythms of these processes are, the closer $\Gamma_\Delta(t)$ will be to 1.

Figures 17 and 18 demonstrate different degrees of coherence for the considered modes. For periodic oscillations (a), both the slow and fast modes of the interacting nephrons are perfectly locked during the observation time. For a system with complex oscillations subjected to noise one can speak about a certain degree of synchronization if the periods of locking become significant compared to the characteristic periods of oscillations [41]. Fully incoherent behavior with respect to both oscillatory modes can be observed in (b). In some cases (c,d) we can diagnose synchronization of the slow motions for relatively long time intervals where the frequencies remain almost equal. Fast motions of interacting nephrons can demonstrate

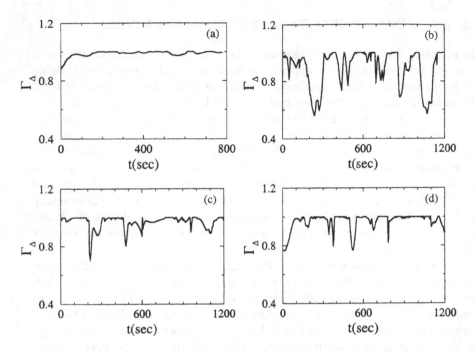

Figure 17.   Mutual wavelet analysis for the slow oscillations. $\Gamma_\Delta = 1$ represents a synchronized state.

different coherence properties. They can be locked during long periods of time together with the slow oscillations (c). We define this type of synchronization as full synchronization since all time scales of the system are locked. Another case (d) is when the fast oscillations are incoherent while the slow oscillations are synchronized during the considered time interval. We refer to this phenomenon as partial synchronization.

The individual oscillatory system has two modes that can be locked with each other. However, an interaction between functional units can break their mutual adjustment. It is also plausible that a coupling will act in different manners on the fast and slow oscillations. For the interacting systems we can introduce two rotation numbers as follows:

$$r_v = T_{v1}/T_{v2}, \quad r_h = T_{h1}/T_{h2} \tag{30}$$

with T_v and T_h denoting the mean return times to appropriately chosen Poincaré sections as defined in Eq. (14). To provide more information, the variation of the phase difference is calculated separately for the slow h and for the fast v oscillations.

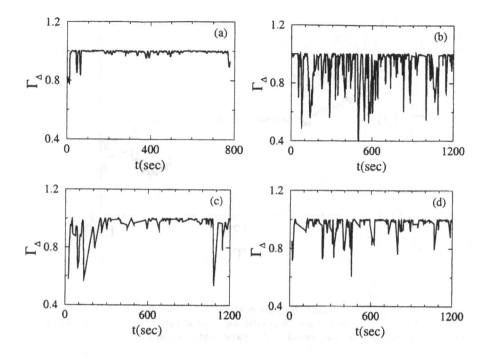

Figure 18.   Mutual wavelet analysis for the fast oscillations. In (d) the fast oscillations are incoherent while the slow oscillations are well-synchronized.

Let us consider the case of $\alpha = 30.0$ corresponding to a weakly developed chaotic attractor in the individual nephron. The coupling strength γ and delay time T_2 in the second nephron are varied. Two different chaotic states can be recognized as asynchronous and synchronous (Fig. 19). For asynchronous behavior the rotation numbers r_h and r_v change continuously with T_2, while inside the synchronization region two cases can be distinguished. To the left ($12.8\ s < T_2 < 13.7\ s$) the rotation numbers r_h and r_v are both equal to unity since both slow and fast oscillations are synchronized. To the right ($T_2 > 13.7\ s$) while the slow h-mode of the chaotic oscillations remains locked, the fast v-mode drifts randomly. In this case the synchronization condition is fulfilled for one of the oscillatory modes only.

5. Conclusion

In this chapter we considered a fully deterministic description both of the function of individual nephron and of the nephron-nephron interaction. With physiologically realistic mechanisms and with independently determined parameters this allowed us to explain how the pressure and flow

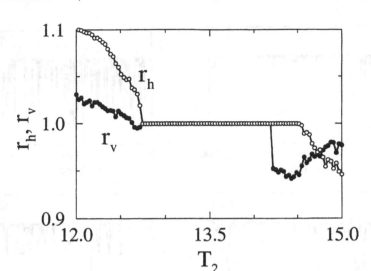

Figure 19. Full and partial synchronization of fast and slow motions ($T_1 = 13.5$ s, $\alpha = 30.0$ and $\gamma = 0.06$). The simulation results depicted in this figure reproduce all the different experimental results observed in the wavelet analysis in Fig. 4.3.

regulation in the nephron becomes unstable in a Hopf bifurcation and how more complicated dynamics can arise as the feedback gain is increased. For coupled nephrons we were able to explain both the observation of in-phase and anti-phase synchronization in the pressure variations for neighboring nephrons in normotensive rats and of chaotic phase synchronization in hypertensive rats. The various synchronization phenomena are likely to have significant physiological ramifications, and transitions between different states of synchronization may play an important role in the regulation of the kidney. In-phase synchronization, for instance, in which the nephrons simultaneously perform the same regulatory adjustments of the incoming blood flow, is likely to produce synergetic effects in the overall response of the system to external disturbances. Out-of-phase synchronization, on the other hand, will produce a slower and less pronounced response of the nephron system in the aggregate.

In practice the nephrons exist and operate in a very noisy environment. The influence of noise is partly illustrated in Fig. 14. Here, the chaotic phase synchronization is interrupted by phase jumps where the synchronization is momentarily lost. A similar effect can arise because the phase oscillators are perturbed by the chaotic (and different) amplitude variations. Noise is also expected to wash out many of details in our bifurcation diagrams, and further investigations obviously have to consider this phenomenon in detail.

Another problem of considerable interest concerns the range of the synchronization between the nephrons. Since the arteriolar network can be mapped out and the lengths and diameters of the various vessels determined, it is possible to obtain an independent estimate of the typical strength of the hemodynamic coupling and its variation across the kidney. Similarly, determination of the decay length for the vascularly propagated signal will allow us to estimate the parameters of that coupling. The typical length of the vascular segments separating neighboring glomeruli is of the order of $250 - 300$ μm. This is only about 30% of the distance that the vascular signal is expected to propagate, suggesting that larger groups of nephrons might act in synchrony. We are presently trying to establish an experimental procedure that will allow us to study the more global aspects of the nephron-nephron interaction. At the same time, we are building a model of an arteriolar tree with a large number of nephrons.

Acknowledgments

This work was partly supported by INTAS (01-2061) and RFBR (01-02-16709). O.S. aknowledges support from the Natural Science Foundation of Denmark.

References

1. L.C. Moore, Tubuloglomerular Feedback and SNGFR Autoregulation in the Rat, *Am. J. Physiol.* **247**, F267-F276 (1984).
2. P.P. Leyssac and L. Baumbach, An Oscillating Intratubular Pressure Response to Alterations in Henle Loop Flow in the Rat Kidney, *Acta Physiol. Scand.* **117**, 415-419 (1983).
3. N.-H. Holstein-Rathlou and P.P. Leyssac, TGF-mediated Oscillations in the Proximal Intratubular Pressure: Differences between Spontaneously Hypertensive Rats and Wistar-Kyoto Rats, *Acta Physiol. Scand.* **126**, 333-339 (1986).
4. P.P. Leyssac and N.-H. Holstein-Rathlou, Effects of Various Transport Inhibitors on Oscillating TGF Pressure Response in the Rat, *Pflügers Archiv* **407**, 285-291 (1986).
5. N.-H. Holstein-Rathlou and D.J. Marsh, Renal Blood Flow Regulation and Arterial Pressure Fluctuations: A Case Study in Nonlinear Dynamics, *Physiol. Rev.* **74**, 637-681 (1994).
6. K.-P. Yip, N.-H. Holstein-Rathlou, and D.J. Marsh, Chaos in Blood Flow Control in Genetic and Renovascular Hypertensive Rats, *Am. J. Physiol.* **261**, F400-F408 (1991).
7. K.-P. Yip and N.-H. Holstein-Rathlou, Chaos and Non-Linear Phenomena in Renal Vascular Control, *Cardiovascular Res.* **31**, 359-370 (1996).

8. K.S. Jensen, N.-H. Holstein-Rathlou, P.P. Leyssac, E. Mosekilde, and D.R. Rasmussen, *Chaos in a System of Interacting Nephrons*, in Chaos in Biological Systems, edited by H. Degn, A.V. Holden, and L.F. Olsen (Plenum, New York, 1987), pp. 23-32.

9. K.S. Jensen, E. Mosekilde, and N.-H. Holstein-Rathlou, Self-Sustained Oscillations and Chaotic Behaviour in Kidney Pressure Regulation, *Mondes en Develop.* **54/55**, 91-109 (1986).

10. E. Mosekilde, *Topics in Nonlinear Dynamics. Applications to Physics, Biology and Economic Systems* (World Scientific, Singapore, 1996).

11. J. Briggs, A Simple Steady-State Model for Feedback Control of Glomerular Filtration Rate, *Kidney Int.* **22**, Suppl. 12, S143-S150 (1982).

12. P.P. Leyssac and N.-H. Holstein-Rathlou, Tubulo-Glomerular Feedback Response: Enhancement in Adult Spontaneously Hypertensive Rats and Effects of Anaesthetics, *Pflügers Archiv* **413**, 267-272 (1989).

13. N.-H. Holstein-Rathlou and D.J. Marsh, A Dynamic Model of the Tubuloglomerular Feedback Mechanism, *Am. J. Physiol.* **258**, F1448-F1459 (1990).

14. N.-H. Holstein-Rathlou and D.J. Marsh, Oscillations of Tubular Pressure, Flow, and Distal Chloride Concentration in Rats, *Am. J. Physiol.* **256**, F1007-F1014 (1989).

15. H.E. Layton, E.B. Pitman and L.C. Moore, Bifurcation Analysis of TGF-Mediated Oscillations in SNGFR, *Am. J. Physiol.* **261**, F904-F919 (1991).

16. N.K. Hollenberg and T. Sandor, Vasomotion of Renal Blood Flow in Essential Hypertension. Oscillations in Xenon Transit, *Hypertension* **6**, 579-585 (1984).

17. N.-H. Holstein-Rathlou, A.J. Wagner, and D.J. Marsh, Tubuloglomerular Feedback Dynamics and Renal Blood Flow Autoregulation in Rats, *Am. J. Physiol.* **260**, F53-F68 (1991).

18. R. Feldberg, M. Colding-Jørgensen, and N.-H. Holstein-Rathlou, Analysis of Interaction between TGF and the Myogenic Response in Renal Blood Flow Autoregulation, *Am. J. Physiol.* **269**, F581-F593 (1995).

19. D. Casellas, M. Dupont, N. Bouriquet, L.C. Moore, A. Artuso and A. Mimran, Anatomic Pairing of Afferent Arterioles and Renin Cell Distribution in Rat Kidneys, *Am. J. Physiol.* **267**, F931-F936 (1994).

20. N.-H. Holstein-Rathlou, Synchronization of Proximal Intratubular Pressure Oscillations: Evidence for Interaction between Nephrons, *Pflügers Archiv* **408**, 438-443 (1987).

21. H. Bohr, K.S. Jensen, T. Petersen, B. Rathjen, E. Mosekilde, and N.-H. Holstein-Rathlou, Parallel Computer Simulation of Nearest-Neighbour Interaction in a System of Nephrons, *Parallel Comp.* **12**, 113-120 (1989).

22. M. Barfred, E. Mosekilde, and N.-H. Holstein-Rathlou, Bifurcation Analysis of Nephron Pressure and Flow Regulation, *Chaos* **6**, 280-287 (1996).

23. M.D. Andersen, N. Carlsson, E. Mosekilde, and N.-H. Holstein-Rathlou, *Dynamic Model of Nephron-Nephron Interaction* in Membrane Transport and Renal Physiology, edited by H. Layton and A. Weinstein (Springer-Verlag, New York, 2002).

24. N.-H. Holstein-Rathlou, K.-P. Yip, O.V. Sosnovtseva, and E. Mosekilde, Synchronization Phenomena in Nephron-Nephron Interaction, *Chaos* **11**, 417-426 (2001).

25. D.E. Postnov, O.V. Sosnovtseva, E. Mosekilde, and N.-H. Holstein-Rathlou, Cooperative Phase Dynamics in Coupled Nephrons, *Int. J. Mod. Phys. B* **15**, 3079-3098 (2001).

26. O.V. Sosnovtseva, D.E. Postnov, E. Mosekilde, and N.-H. Holstein-Rathlou, Syn-

chronization of Tubular Pressure Oscillations in Interacting Nephrons, *Chaos, Solitons and Fractals* (2002) (in press).

27. E. Mosekilde, Yu. Maistrenko, and D. Postnov, *Chaotic Synchronization – Applications to Living Systems* (World Scientific, Singapore, 2002).

28. N.-H. Holstein-Rathlou and P.P. Leyssac, Oscillations in the Proximal Intratubular Pressure: A Mathematical Model, *Am. J. Physiol.* **252**, F560-F572 (1987).

29. W.M. Deen, C.R. Robertson, and B.M. Brenner, A Model of Glomerular Ultrafiltration in the Rat, *Am. J. Physiol.* **223**, 1178-1183 (1984).

30. M. Rosenbaum and D. Race, Frequency-Response Characteristics of Vascular Resistance Vessels, *Am. J. Physiol.* **215**, 1397-1402 (1968).

31. Y.-C. B. Fung, *Biomechanics. Mechanical Properties of Living Tissues* (Springer, New York, 1981).

32. T.S. Parker and L.O. Chua, *Practical Numerical Algorithms for Chaotic Systems* (Springer-Verlag, Berlin, 1989).

33. M. Marek and I. Schreiber, *Chaotic Behavior in Deterministic Dissipative Systems* (Cambridge University Press, England, 1991).

34. A. Wolf, *Quantifying Chaos with Lyapunov Exponents* in Chaos, edited by A.V. Holden (Manchester University Press, England, 1986).

35. Y.-M. Chen, K.-P. Yip, D.J. Marsh and N.-H. Holstein-Rathlou, Magnitude of TGF-Initiated Nephron-Nephron Interactions is Increased in SHR, *Am. J. Physiol.* **269**, F198-F204 (1995).

36. V.I. Arnol'd, Small Denominators. I. Mapping of the Circumference onto Itself, *Am. Math. Soc. Transl.* Ser. 2 **46**, 213-284 (1965).

37. K.-P. Yip, N.-H. Holstein-Rathlou, and D.J. Marsh, Dynamics of TGF-Initiated Nephron-Nephron Interactions in Normotensive rats and SHR, *Am. J. Physiol.* **262**, F980-F988 (1992).

38. Ö. Källskog and D.J. Marsh, TGF-Initiated Vascular Interactions between Adjacent Nephrons in the Rat Kidney, *Am. J. Physiol.* **259**, F60-F64 (1990).

39. M.G. Rosenblum, A.S. Pikovsky, and J. Kurths, Phase Synchronization of Chaotic Oscillators, *Phys. Rev. Lett.* **76**, 1804-1807 (1996).

40. A.S. Pikovsky, M.G. Rosenblum, G.V. Osipov, and J. Kurths, Phase Synchronization of Chaotic Oscillators by External Driving, *Physica D* **104** 219-238 (1997).

41. L.R. Stratonovich, *Topics in the Theory of Random Noise* (Gordon and Breach, New York, 1963).

42. D.E. Postnov, T.E. Vadivasova, O.V. Sosnovtseva, A.G. Balanov, V.S. Anishchenko, and E. Mosekilde, Role of Multistability in the Transition to Chaotic Phase Synchronization, *Chaos* **9**, 227-232 (1999).

43. T.E. Vadivasova, O.V. Sosnovtseva, A.G. Balanov, and V.V. Astakhov, Phase Multistability of Synchronous Chaotic Oscillations, *Discrete Dynamics in Nature and Society* **4**, 231-243 (2000).

44. J. Rasmusen, E. Mosekilde, and Chr. Reick, Bifurcations in Two Coupled Rössler Systems, *Math. Comp. Sim.* **40**, 247-270 (1996).

45. O.V. Sosnovtseva, D.E. Postnov, A.M. Nekrasov, and E. Mosekilde, Phase Multistability Analysis of Self-Modulated Oscillations, *Phys. Rev. E* (2002) (in press).

46. A. Grossmann, J. Morlet, S.I.A.M. J. Math. Anal. **15**, 723 (1984); I. Daubechies, *Ten Lectures on Wavelets* (S.I.A.M., Philadelphia, 1992).

47. Y. Kuramoto, *Chemical Oscillations, Waves, and Turbulence* (Springer-Verlag, New York, 1984).

48.   S.H. Park, S. Kim, H.-B. Pyo, and S. Lee, Multistability analysis of phase locking patterns in an excitatory coupled neural system, *Phys. Rev. E* **60**, 2177-2181 (1999).
49.   S.K. Han, C. Kurrer, and Y. Kuramoto, Dephasing and bursting in coupled neural oscillators, *Phys. Rev. Lett.* **75** 3190-3193 (1995).
50.   D. Postnov, S.K. Han, and H. Kook, Synchronization of diffusively coupled oscillators near the homoclinic bifurcation, *Phys. Rev. E* **60**, 2799-2807 (1999).
51.   N.-H. Holstein-Rathlou, Synchronization of proximal intratubular pressure oscillation: evidence for interaction between nephrons *Pflügers Archiv* **408**, 438-443 (1987).

SYNCHRONY IN GLOBALLY COUPLED CHAOTIC, PERIODIC, AND MIXED ENSEMBLES OF DYNAMICAL UNITS

EDWARD OTT[1], PAUL SO[2], ERNEST BARRETO[2] and THOMAS ANTONSEN[1]

[1] *Institute for Research in Electronics and Applied Physics, Department of Physics, and Department of Electrical and Computer Engineering, University of Maryland, College Park, Maryland, 20742*

[2] *Department of Physics and Astronomy and the Krasnow Institute for Advanced Study, George Mason University, Fairfax, Virginia, 22030*

Abstract

The onset of collective synchronous behavior in globally coupled ensembles of oscillators is discussed. We present a formalism that is applicable to general ensembles of heterogeneous, continuous time dynamical units that, when uncoupled, are chaotic, periodic, or a mixture of both. A discussion of convergence issues, important for the proper implementation of our method, is included. Our work leads to a quantitative prediction for the critical coupling value at the onset of collective synchrony and for the growth rate of the resulting coherent state.

Systems that consist of many coupled heterogeneous dynamical units are of great interest in a wide variety of situations. The simplest such situation is where the coupling is *global* in the sense that each unit is coupled to all other units. Past work has concentrated on the case where the dynamics of the uncoupled units is periodic with a spread in the oscillator frequencies [1]. In that case, a typical behavior is that, for sufficiently low coupling, the individual units oscillate incoherently, but that, as the magnitude of the coupling increases through a critical value, there is a transition to coherent system dynamics in which a group of oscillators in the ensemble of units becomes locked in frequency and phase. Possible applications include synchrony in chirping crickets [2], flashing fireflies [3], Josephson junction

175

A. Pikovsky and Y. Maistrenko (eds.), Synchronization: Theory and Application, 175–186.
© 2003 Kluwer Academic Publishers. Printed in the Netherlands.

arrays [4], semiconductor laser arrays [5], and cardiac pacemakers cells [6]. More recently, the similar question of what happens when the uncoupled dynamics of the individual units is chaotic has been addressed [7, 8, 9] (see also the recent experimental study of globally coupled chaotic electrochemical oscillators [10]). In this contribution we discuss a formalism that is capable of treating the onset of synchronism of a general system of globally coupled, heterogeneous, continuous-time dynamical units. No *a priori* assumption regarding the uncoupled dynamics of the individual units is made. Thus, one can consider chaotic or periodic dynamics of the uncoupled units, including the case where both types of units are present in the same system.

In order to illustrate the context for the subsequent presentation of our analysis, we first discuss some numerical examples. We consider a system consisting of an ensemble of units in which each one of the units, when uncoupled from the others, obeys the Lorenz equations: $dx^{(1)}/dt = \sigma(x^{(2)} - x^{(1)})$; $dx^{(2)}/dt = rx^{(1)} - x^{(2)} - x^{(1)}x^{(3)}$; $dx^{(3)}/dt = -bx^{(3)} + x^{(1)}x^{(2)}$. We denote the state variables for unit i of the ensemble by $(x_i^{(1)}, x_i^{(2)}, x_i^{(3)})$ and we couple the units by adding a global coupling term $(K/N)\sum_{i=1}^{N} x_i^{(1)}(t)$ to the right side of the $dx_i^{(1)}/dt$ equation. Here, N is the number of units in the ensemble and is presumed to be large ($N >> 1$); in the subsequent analysis, the limit $N \to \infty$ is employed. In addition, the parameter r will be different for each ensemble member, and we denote its value for unit i by r_i. For our numerical examples, the values of r_i are taken to be uniformly distributed in an interval $[r_-, r_+]$ and $N > 10^4$. We numerically examine the following three ensembles:

Case (a): chaotic ensemble, $[r_-, r_+] = [28, 52]$.
Case (b): periodic ensemble, $[r_-, r_+] = [150, 165]$.
Case (c): mixed ensemble, $[r_-, r_+] = [167, 202]$.

In case (a), numerical computations show that the uncoupled Lorenz equations yield chaotic solutions with no discernable windows of periodicity in the range $[r_-, r_+]$. In case (b), the uncoupled solutions are periodic, but there is a pitchfork bifurcation at a value $r = r_p$, $r_- < r_p < r_+$ such that as r decreases through r_p, a symmetric periodic orbit where the symmetry is $(x_i^{(1)}, x_i^{(2)}, x_i^{(3)}) \to (-x_i^{(1)}, -x_i^{(2)}, x_i^{(3)})$ bifurcates to two individually asymmetric orbits. In case (c), the ensemble is substantially chaotic but with a large window of periodicity in $[r_-, r_+]$. Figures viii (a-c) give numerical results for cases (a-c). These figures show an order parameter $\bar{x}_T$ versus the coupling coefficient k, where

$$\bar{x}_T = \left\{ \frac{1}{T} \int_t^{t+T} \left(\frac{1}{N} \sum_{i=1}^{N} x_i^{(1)}(t') \right)^2 dt' \right\}^{1/2}$$

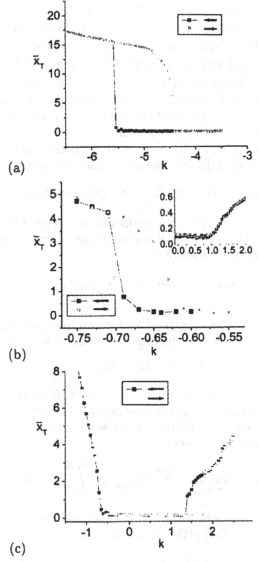

Figure 1. $\bar{x}_T$ versus k for cases (a), (b), and (c).

and $\bar{x}_T$ characterizes the degree of coherent motion of the system. The times t and T are chosen large enough that the system settles into its time-asymptotic dynamics and $\bar{x}_T$ is essentially independent of T. Note that, if the individual units behave incoherently, the sum is close to zero by the symmetry of the uncoupled Lorenz equations. For case (a), there is a subcritical bifurcation as k decreases through $k_a = -5.6$. For case (b), there is a supercritical Hopf bifurcation as k increases through $k_b^{(+)} \approx 1$ and a subcritical Hopf bifurcation as k decreases through $k^{(-)} \approx -0.68$. For case (c), there is a subcritical Hopf bifurcation at $k_c^{(+)} \approx 1.8$, and a supercritical Hopf bifurcation at $K_c^{(-)} \approx -0.7$. It is our goal to obtain a theory for these critical k values at the onset of coherence. These mark the onset of instability of the incoherent state, and we are also interested in the exponential growth rates of these instabilities.

We now present our analysis, treating the simplest case (generalizations will be given elsewhere [11]). We consider dynamical systems of the form

$$dx_i(t)/dt = \mathbf{G}(\mathbf{x}_i(t), \mathbf{\Omega}_i) + \mathbf{K}(\langle\langle\mathbf{x}\rangle\rangle_* - \langle\langle\mathbf{x}(t)\rangle\rangle), \tag{1}$$

where $\mathbf{x}_i = (x_i^{(1)}, x_i^{(2)}, \ldots, x_i^{(q)})^T$ is a q-dimensional vector; $\mathbf{G}$ is a q-dimensional vector function; $\mathbf{K}$ is a constant $q \times q$ coupling matrix; $i = 1, 2, \cdots, N$; $\langle\langle\mathbf{x}(t)\rangle\rangle$ is the instantaneous average

$$\langle\langle\mathbf{x}(t)\rangle\rangle = \lim_{N\to\infty} N^{-1} \sum_i \langle\mathbf{x}_i(t)\rangle, \tag{2}$$

and, for each i, $\langle\mathbf{x}_i\rangle$ is the average of $\mathbf{x}_i$ over an infinite number of initial conditions $\mathbf{x}_i(0)$ distributed on the attractor of the ith uncoupled system,

$$d\mathbf{x}_i/dt = \mathbf{G}(\mathbf{x}_i, \mathbf{\Omega}_i). \tag{3}$$

$\mathbf{\Omega}_i$ is a parameter vector specifying the uncoupled ($\mathbf{K} = 0$) dynamics, and $\langle\langle\mathbf{x}\rangle\rangle_*$ is the *natural measure* [12] and i average of the state of the uncoupled system. That is, to compute $\langle\langle\mathbf{x}\rangle\rangle_*$, we set $\mathbf{K} = 0$, compute the solutions to Eq. (3), and obtain $\langle\langle\mathbf{x}\rangle\rangle_*$ from

$$\langle\langle\mathbf{x}\rangle\rangle_* = \lim_{N\to\infty} N^{-1} \sum_i \left[\lim_{\tau_0\to\infty} \tau_0^{-1} \int_0^{\tau_0} \mathbf{x}_i(t)dt \right]. \tag{4}$$

In what follows we assume that the $\mathbf{\Omega}_i$ are randomly chosen from a smooth probability density function $\rho(\mathbf{\Omega})$. Thus, an alternate means of expressing (4) is

$$\langle\langle\mathbf{x}\rangle\rangle_* = \int \mathbf{x}\rho(\mathbf{\Omega})d\mu_\mathbf{\Omega} d\mathbf{\Omega}, \tag{5}$$

where $\mu_\mathbf{\Omega}$ is the natural invariant measure for the system $d\mathbf{x}/dt = \mathbf{G}(\mathbf{x}, \mathbf{\Omega})$. By construction, $\langle\langle\mathbf{x}\rangle\rangle = \langle\langle\mathbf{x}\rangle\rangle_*$ is a solution of the globally coupled system (1). We call this solution the "incoherent state" because the coupling term cancels and the individual oscillators do not affect each other. The question we address is whether the incoherent state is stable. In particular, as a system parameter such as the coupling strength varies, the onset of instability of the incoherent state signals the start of coherent, synchronous behavior of the ensemble.

To perform the stability analysis, we assume that the system is in the incoherent state, so that at any fixed time t, and for each i, $\mathbf{x}_i(t)$ is distributed according to the natural measure. We then perturb the orbits $\mathbf{x}_i(t) \rightarrow \mathbf{x}_i(t) + \delta\mathbf{x}_i(t)$, where $\delta\mathbf{x}_i(t)$ is an infinitesimal perturbation:

$$d\delta\mathbf{x}_i/dt = \mathbf{DG}(\mathbf{x}_i(t), \mathbf{\Omega}_i)\delta\mathbf{x}_i - \mathbf{K}\langle\langle\delta\mathbf{x}_i\rangle\rangle \tag{6}$$

where

$$\mathbf{DG}(\mathbf{x}_i(t), \mathbf{\Omega}_i)\delta\mathbf{x}_i = \delta\mathbf{x}_i \cdot \frac{\partial}{\partial\mathbf{x}_i}\mathbf{G}(\mathbf{x}_i(t), \mathbf{\Omega}_i).$$

Introducing the fundamental matrix $\mathbf{M}_i(t)$ for system (6),

$$d\mathbf{M}_i/dt = \mathbf{DG} \cdot \mathbf{M}_i, \tag{7}$$

where $\mathbf{M}_i(0) \equiv \mathbb{1}$, we can write the solution of Eq. (6) as

$$\delta\mathbf{x}_i(t) = -\int_{-\infty}^{t} \mathbf{M}_i(t)\mathbf{M}_i^{-1}(\tau)\mathbf{K}\langle\langle\delta\mathbf{x}\rangle\rangle_\tau d\tau, \tag{8}$$

where we use the notation $\langle\langle\delta\mathbf{x}\rangle\rangle_\tau$ to signify that $\langle\langle\delta\mathbf{x}\rangle\rangle$ is evaluated at time τ. Note that, through Eq. (7), $\mathbf{M}_i$ depends on the unperturbed orbits $\mathbf{x}_i(t)$ of the uncoupled nonlinear system (3), which are determined by their initial conditions $\mathbf{x}_i(0)$ (distributed according to the natural measure).

Assuming that $\langle\langle\delta\mathbf{x}\rangle\rangle$ evolves exponentially in time (i.e., $\langle\langle\delta\mathbf{x}\rangle\rangle = \mathbf{\Delta}e^{st}$), Eq. (8) yields

$$\{\mathbb{1} + \tilde{\mathbf{M}}(s)\mathbf{K}\}\mathbf{\Delta} = 0, \tag{9}$$

where s is complex, and

$$\tilde{\mathbf{M}}(s) = \left\langle\!\!\left\langle \int_{-\infty}^{t} e^{-s(t-\tau)}\mathbf{M}_i(t)\mathbf{M}_i^{-1}(\tau)d\tau \right\rangle\!\!\right\rangle_*. \tag{10}$$

Thus the dispersion function determining s is

$$D(s) = \det\{\mathbb{1} + \tilde{\mathbf{M}}(s)\mathbf{K}\} = 0. \tag{11}$$

In order for Eqs. (9) and (11) to make sense, the right side of Eq. (10) must be independent of time. As written, it may not be clear that this is so. We now demonstrate this, and express $\tilde{\mathbf{M}}(s)$ in a more convenient form. Writing the dependence of $\mathbf{M}_i$ in Eq. (10) on the initial condition explicitly, we have from the definition of $\mathbf{M}_i$, $\mathbf{M}_i(t, \mathbf{x}_i(0))\mathbf{M}_i^{-1}(\tau, \mathbf{x}_i(0)) = \mathbf{M}_i(t-\tau, \mathbf{x}_i(\tau)) = \mathbf{M}_i(T, \mathbf{x}_i(t-T))$, where $T = t-\tau$. Using this in Eq. (10), we have

$$\tilde{\mathbf{M}}(s) = \langle\langle \int_{0}^{\infty} e^{-sT}\mathbf{M}_i(T, \mathbf{x}_i(t-T))dT\rangle\rangle_*.$$

Note that our solution requires that the integral in the above converge. Since the growth of $\mathbf{M}_i$ with increasing T is dominated by h_i, the largest Lyapunov exponent for the orbit $\mathbf{x}_i$, we require $Re(s) > \Gamma$, where $\Gamma = \max_{\mathbf{x}_i, \Omega_i}(h_i)$. In contrast with the chaotic case where $\Gamma > 0$, an ensemble of periodic attractors has $\Gamma = 0$ (for an attracting periodic orbit $h_i = 0$ corresponds to orbit perturbations along the flow). With the condition $Re(s) > \Gamma$, the integral converges exponentially and uniformly in the quantities over which we average. Thus we can interchange the integration and the average,

$$\tilde{\mathbf{M}}(s) = \int_0^\infty e^{-sT} \langle\langle \mathbf{M}_i(T, \mathbf{x}_i(t - T)) \rangle\rangle_* dT. \tag{12}$$

In Eq. (12) the only dependence on t is through the initial condition $\mathbf{x}_i(T - t)$. However, since the quantity within angle brackets includes not only an average over i, but also an average over initial conditions with respect to the natural measure of each uncoupled attractor i, the time invariance of the natural measure ensures that Eq. (12) is independent of t. In particular, invariance of a measure means that if an infinite cloud of initial conditions $\mathbf{x}_i(0)$ is distributed on uncoupled attractor i at $t = 0$ according to its natural invariant measure, then the distribution of the orbits, as they evolve to any time t via the uncoupled dynamics (Eq. (3)), continues to give the same distribution as at time $t = 0$. Hence, although $\mathbf{M}_i(T, \mathbf{x}_i(t - T))$ depends on t, when we average over initial conditions, the result $\langle \mathbf{M}_i(T, \mathbf{x}_i(t - T)) \rangle_*$ is independent of t for each i. Thus we drop the dependence of $\langle\langle \mathbf{M}_i \rangle\rangle_*$ on the initial values of the $\mathbf{x}_i$ and write

$$\tilde{\mathbf{M}}(s) = \int_0^\infty e^{-sT} \langle\langle \mathbf{M}(T) \rangle\rangle_* dT, \tag{13}$$

where, for convenience we have also dropped the subscript i. Thus $\tilde{\mathbf{M}}$ is the Laplace transform of $\langle\langle \mathbf{M} \rangle\rangle_*$. As we will see, this result for $\tilde{\mathbf{M}}(s)$ can be analytically continued into $Re(s) < \Gamma$.

Note that $\tilde{\mathbf{M}}(s)$ depends only on the solution of the linearized *uncoupled* system (Eq. (7)). Hence the utility of the dispersion function $D(s)$ given by Eq. (11) is that it determines the linearized dynamics of the globally coupled system in terms of those of the individual uncoupled systems.

Consider the jth column of $\langle\langle \mathbf{M}(t) \rangle\rangle_*$, which we denote $[\langle\langle \mathbf{M}(t) \rangle\rangle_*]_j$. According to our definition of $\mathbf{M}_i$ given by Eq. (7), we can interpret $[\langle\langle \mathbf{M}(t) \rangle\rangle_*]_j$ as follows. Assume that for each of the uncoupled systems i in Eq. (3), we have a cloud of an infinite number of initial conditions sprinkled randomly according to the natural measure on the uncoupled attractor. Then, at $t = 0$, we apply an equal infinitesimal displacement δ_j in the direction j to each orbit in the cloud. That is, we replace $\mathbf{x}_i(0)$ by $\mathbf{x}_i(0) + \delta_j \mathbf{a}_j$, where $\mathbf{a}_j$ is a unit vector in $\mathbf{x}$-space in the direction j. Since the particle cloud is displaced from the attractor, it relaxes back to the attractor as time evolves. The quantity $[\langle\langle\mathbf{M}\rangle\rangle_*]_j\delta_j$ gives the time evolution of the i-averaged perturbation of the centroid of the cloud as it evolves back to the attractor and redistributes itself on the attractor.

We now argue that $\langle\langle\mathbf{M}\rangle\rangle_*$ decays to zero exponentially with increasing time. We consider the general case where the support of the smooth density $\rho(\Omega)$ contains open regions of Ω for which the dynamical system (3) has attracting periodic orbits as well as a positive measure of Ω on which Eq. (3) has chaotic orbits. Numerical experiments on chaotic attractors (including structurally unstable attractors) generally show that they are strongly mixing; i.e., a cloud of many particles rapidly arranges itself on the attractor according to the natural measure. Thus, for each Ω_i giving a chaotic attractor, it is reasonable to assume that the average of $\mathbf{M}_i$ over initial conditions $\mathbf{x}_i(0)$, denoted $\langle\mathbf{M}_i\rangle_*$, decays exponentially. For a periodic attractor, however, $\langle\mathbf{M}_i\rangle_*$ does not decay: a distribution of orbits along a limit cycle comes to the same distribution after one period, and this repeats forever. Thus, if the distribution on the limit cycle was noninvariant, it remains noninvariant and oscillates forever at the period of the periodic orbit. On the other hand, periodic orbits exist in open regions of Ω, and, when we average over Ω, there is the possibility that with increasing time cancellation causing decay occurs via the process of "phase mixing". For this case we appeal to an example. In particular, the explicit computation of $\langle\mathbf{M}_i\rangle_*$ for a simple model limit cycle ensemble is given in Ref. [11]. The result is

$$\langle\mathbf{M}_i\rangle_* = \frac{1}{2}\left[\begin{array}{cc} \cos\Omega_i t & -\sin\Omega_i t \\ \sin\Omega_i t & \cos\Omega_i t \end{array}\right],$$

and indeed this oscillates and does not decay to zero. However, if we average over the oscillator distribution $\rho(\Omega)$ we obtain

$$\langle\langle\tilde{\mathbf{M}}\rangle\rangle_* = \frac{1}{2}\left[\begin{array}{cc} c(t) & -s(t) \\ s(t) & c(t) \end{array}\right],$$

where $c(t) = \int \rho(\Omega)\cos\Omega t d\Omega$ and $s(t) = \int \rho(\Omega)\sin\Omega t d\Omega$. For any analytic $\rho(\Omega)$ these integrals decay exponentially with time. Thus, based on these considerations of chaotic and periodic attractors, we see that for sufficiently smooth $\rho(\Omega)$, there is reason to believe that $\langle\langle\mathbf{M}\rangle\rangle_*$, the average of $\mathbf{M}_i$ over $\mathbf{x}_i(0)$ and over Ω_i, decays exponentially to zero with increasing time. Conjecturing this decay to be exponential, $\|\langle\langle\mathbf{M}(t)\rangle\rangle_*\| < \kappa e^{-\gamma t}$ for positive constants κ and γ, we see that the integral in Eq. (13) converges for $Re(s) > -\gamma$. This conjecture is supported by our numerical results. Thus, while Eq. (13) was derived under the assumption $Re(s) > \Gamma > 0$, using analytic continuation, we can regard Eq. (13) as valid for $Re(s) > -\gamma$. Note that, for our purposes, it suffices to require only that $\|\langle\langle \mathbf{M}(t)\rangle\rangle_*\|$ be bounded, rather than that it decay exponentially. Boundedness corresponds to $\gamma = 0$, which is enough for us, since, as soon as instability occurs, the relevant root of $D(s)$ has $Re(s) > 0$.

In order to apply Eq. (11), to a given situation, it is necessary to numerically approximate the matrix $\tilde{\mathbf{M}}(s)$. To do this we consider two possible candidate approaches.

Approach (i): First approximate the natural measure on each attractor i by a large finite number of orbits initially distributed according to the natural measure. For each initial condition, obtain $\mathbf{x}_i(t)$ from Eq. (3). Use these solutions in $\mathbf{DG}$ and solve Eq. (7). Then average over the natural measure and i to obtain $\langle\langle \mathbf{M}(t)\rangle\rangle_*$, and do the Laplace transform (Eq. (13).

Approach (ii): Since $\langle\langle \mathbf{M}\rangle\rangle_*$ is the response to an impulse (i.e., the sudden displacement of each orbit), its Laplace transform multiplied by $\exp(st)$, namely $M(s)\exp(st)$, is the response to the drive $\exp(st)\mathbf{1}$ added to the right side of Eq. (6). This suggests the following numerical prodedure for finding M(s). Solve

$$\frac{d\tilde{\mathbf{x}}_i^{(c,s)}}{dt} = \mathbf{G}(\tilde{\mathbf{x}}_i^{(c,s)}, \mathbf{\Omega}_i) + \mathbf{\Delta}_j a_j e^{\sigma t} \begin{cases} \cos \omega t \\ \sin \omega t, \end{cases} \tag{14}$$

where $s = \sigma - i\omega$, and $\mathbf{a}_j$ is a unit vector in the direction j. For large t, but $\delta_j \exp(\sigma t)$ still small throughout the time interval $(0, t)$, we can regard the average response as approximately linear. Thus, the jth column of $\tilde{\mathbf{M}}(s)$ is

$$[\langle\langle \mathbf{M}(t)\rangle\rangle_*]_j = \mathbf{\Delta}_j^{-1}(\langle\langle \mathbf{x}\rangle\rangle_* - \langle\langle \mathbf{x}'\rangle\rangle), \tag{15}$$

where $\tilde{\mathbf{x}}_i = \tilde{\mathbf{x}}_i^{(c)} - \tilde{\mathbf{x}}_i^{(s)}$. Numerically, $\langle\langle \tilde{\mathbf{x}}\rangle\rangle$ can be approximated using a large finite number of orbits. In Ref. [8], a technique equivalent to this with s taken to be imaginary ($s = -i\omega$) was used to obtain marginal stability [13].

For the coupling we have chosen to use for our numerical experiments, only $\tilde{M}_{11}(s)$ is nonzero, and thus Eq. (11) reduces to

$$1 + \tilde{M}_{11}(-i\omega)k = 0 \tag{16}$$

where we have set $s = -i\omega$. Solving $\text{Im}[\tilde{M}_{11}(-i\omega)] = 0$ yields roots $\omega = \omega^*$, which, when reinserted into Eq. (16) yield possible values $k = k^* = -[\tilde{M}_{11}(-i\omega^*)]^{-1}$ for the critical coupling strengths. To determine which of the possibly multiple roots ω^* are relevant, we envision that as k is increased or decreased from zero, a critical coupling value is encountered at which the incoherent state first becomes unstable. Hence we are interested in the roots $\omega_{a,b}^*$ corresponding to the smallest $|k^*|$ for k^* both negative

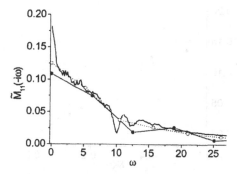

Figure 2. $\tilde{M}_{11}(-i\omega)$ versus ω for case (a). The solid black line is $\mathrm{Re}(\tilde{M}_{11})$ from approach (ii); the solid grey line is $\mathrm{Im}(\tilde{M}_{11})$ from approach (ii); and the dashed line is $\mathrm{Re}(\tilde{M}_{11})$ from approach (i).

$(k^* = -|k_b^*|)$ and positive $(k^* = k_b^* > 0)$. Growth rates and frequency shifts from ω^* can also be simply obtained for k near k^* by setting $k = k^* + \delta k$, $s = -i(\omega^* + \delta\omega) + \gamma$ and expanding Eq. (11) for small δk, $\delta\omega$, and γ; e.g.,

$$\gamma = -\frac{\delta k}{(k^*)^2}\frac{\partial Im[\tilde{M}_{11}(-i\omega)]/\partial\omega}{|\partial\tilde{M}_{11}(-i\omega)/\partial\omega|^2} \tag{17}$$

where $\partial\tilde{M}_{11}/\partial\omega$ is evaluated at $\omega = \omega^*$.

We now illustrate the above by application to our Lorenz ensemble, case (a) (see Fig. viii(a)). Related results for cases (b) and (c) will be reported elsewhere [11]. The black and grey solid lines in Fig. viii show $\mathrm{Re}[\tilde{M}_{11}(-i\omega)]$ and $\mathrm{Im}[\tilde{M}_{11}(-i\omega)]$ versus ω as obtained using approach (ii) with $\Delta_x = 2$ and $N = 20,000$. (We also tested other values of Δ_x up to 5, obtaining similar results, thus indicating that the perturbation is sufficiently linear.) $\mathrm{Im}[\tilde{M}_{11}(-i\omega)]$ crosses zero only at $\omega^* = 0$, where $\mathrm{Re}[\tilde{M}_{11}(-i\omega)]$ has a prominent peak. This gives a critical coupling value of -5.6 ± 0.15 in reasonable agreement with the threshold for coherence observed in Fig. viii(a). Figure viii shows the instability growth rate from Eq. (17) versus δk as a solid line, along with values observed from simulations of the full nonlinear system plotted as dots. To obtain the latter data, we first initialize the ensemble in the incoherent state by time evolution with the coupling k set to zero. We then turn on the coupling $k = -|k^*| + \delta k$, plot $\ln\langle\langle x^{(1)}\rangle\rangle$ versus t, and fit a straight line to the resulting graph during the exponential growth phase. As can be seen from Fig. viii, the result obtained from Eq. (17) agrees well with the data for $0 \geq \delta k \geq -0.6$.

Figure viii shows the result of a computation of $\langle\langle M_{11}(t)\rangle\rangle_*$ versus t by the use of approach (i) with $N = 20,000$. $\langle\langle M_{11}(t)\rangle\rangle_*$ behaves as expected for $t \leq 0.7$ (i.e., it decays with time), but past that time it shows apparent

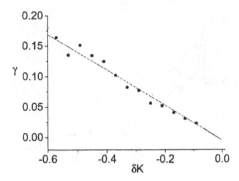

Figure 3. The growth rate γ versus δk for case (a).

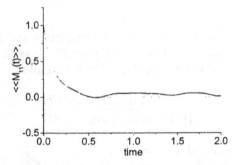

Figure 4. $\langle\langle M_{11}(t)\rangle\rangle_*$ from approach (i) versus t, $N = 20,000$.

divergent behavior. This can be understood on the basis that the individual $\mathbf{M}_i(t)$ for each orbit diverge exponentially at their largest Lyapunov exponent. By our previous arguments, however, we know that the average $\langle\langle M_{11}(t)\rangle\rangle_*$ decays. Thus the average must result in cancellation of the exponential growth components. However, since $\langle\langle M_{11}(t)\rangle\rangle_*$ decays exponentially, and the individual $\mathbf{M}_i(t)$ grow exponentially, this cancellation becomes more and more delicate as time increases. Thus, for any finite N, divergence of the method will always occur at large time. The question is whether a believable result can be obtained for a time duration that is long enough to be useful. Calculating $\mathrm{Re}[\tilde{M}_{11}(-i\omega)]$ from the result in Fig. viii by doing the Laplace transform only over the reliable range $0 \leq t \leq 0.7$, we obtain the result shown in Fig. viii. While there is reasonable agreement with the result from approach (ii) for $\omega \geq 0.1$, approach (i) fails to capture the important sharp increase to the peak at $\omega = 0$ which occurs for for $\omega \leq 0.1$. The reason for this is that this feature would correspond to a time scale $1/\omega \sim 10$ which is well past the finite N-induced divergence in Fig. viii. Thus approach (i) yields a value of $|k^*|$ that is too large (by a factor or order 2). While approach (i) fails in this case, it can be useful in other cases depending on the strength of the divergences that the system exhibits, and particularly in periodic ensembles (e.g., case (b)) where M_i does not grow exponentally.

In conclusion, we have presented a general formulation for the determination of the stability of the incoherent state of a globally coupled system of continuous time dynamical units. The formalism is valid for both chaotic and periodic dynamics of the individual units. We discuss the analytic properties of $\tilde{M}(s)$ and its numerical determination. We find that these are connected: analytic continuation of $\tilde{M}(s)$ to the $Im(s)$ axis is necessary for application of the analysis, but in the chaotic case, can lead to numerical difficulties in determining $\tilde{M}(s)$ (Fig. viii). Our numerical examples illustrate the validity of the approach, as well as practical limitation to numerical application.

This work was supported by grants from ONR (physics), NSF (PHYS-0098632 and IBN9727739), and NIH (K25MH01963).

References

1. Y. Kuramoto, in *International Symposium on Mathematical Problems in Theoretical Physics*, edited by H. Araki, Lecture Notes in Physics, Vol. 39 (Springer, Berlin, 1975); *Chemical Oscillators, Waves and Turbulence* (Springer, Berlin, 1984); A. T. Winfree, *The Geometry of Biological Time* (Springer, New York, 1980); for a review of work on the Kuramoto model, see S. H. Strogatz, Physica D **143**, 1 (2000).
2. T. J. Walker, Science **166**, 891 (1969).
3. J. Buck, Q. Rev. Biol. **63**, 265 (1988).
4. K. Wiesenfeld, P. Colet and S. H. Strogatz, Phys. Rev. Lett. **76**, 404 (1996).
5. D. V. Ramana Reddy, A. Sen and G. L. Johnston, Phys. Rev. Lett. **80**, 5109 (1998).
6. C. S. Peskin, *Mathematical Aspects of Heart Physiology* (Courant Institute of Mathematical Sciences, New York, 1975).
7. A. S. Pikovsky, M. G. Rosenblum and J. Kurths, Europhys. Lett. **34**, 165 (1996).
8. H. Sakaguchi, Phys. Rev. E **61**, 7212 (2000).
9. D. Topaj, W. -H. Kye and A. Pikovsky, Phys. Rev. Lett. **87**, 074101 (2001).
10. I. Kiss, Y. Zhai and J.L. Hudson, Phys. Rev. Lett. **88**, 238301 (2002).
11. E. Ott, P. So, E. Barreto, and T.M. Antonsen, nlin.CD/0205018.
12. E. Ott, *Chaos in Dynamical Systems* (Cambridge Univ. Press, 1993), Chapter 3. For a given Ω, the natural measure μ_Ω for an attractor A of the uncoupled system $dx/dt = G(x, \Omega)$ gives the fraction of time $\mu_\Omega(S)$ that a *typical* infinitely long orbit originating in $B(A)$ (the basin of attraction of A) spends in a subset S of state space. By the word typical we refer to the supposition that there is a set of initial conditions $x(0)$ in $B(A)$ where this set has Lebesgue measure (roughly volume) equal to the Lebesgue measure of $B(A)$ and such that each initial condition in this set gives the same value (i.e., the natural measure) for the fraction of time spent in S by the resulting orbit.
13. Other methods for calculating $\tilde{M}(s)$ are also possible (see [11]). In particular, in the case of a chaotic ensemble, a technique based on unstable periodic orbits embedded

in the chaotic attractor in conjunction with cycle expansions appears to be attractive.
This approach is presently under investigation.

PHASE SYNCHRONIZATION OF REGULAR AND CHAOTIC SELF-SUSTAINED OSCILLATORS

ARKADY S. PIKOVSKY and MICHAEL G. ROSENBLUM
Department of Physics, Potsdam University, Am Neuen Palais 19, PF 601553, D-14415, Potsdam, Germany

Abstract

In this review article we discuss effects of phase synchronization of nonlinear self-sustained oscillators. Starting with a classical theory of phase locking, we extend the notion of phase to autonoumous continuous-time *chaotic* systems. Using as examples the well-known Lorenz and Rössler oscillators, we describe the phase synchronization of chaotic oscillators by periodic external force. Both statistical and topological aspects of this phenomenon are discussed. Then we proceed to more complex cases and discuss phase synchronization in coupled systems, lattices, large globally coupled ensembles, and of space-time chaos. Finally, we demonstrate how the synchronization effects can be detected from observations of real data.

1. Introduction

Synchronization, a basic nonlinear phenomenon, discovered at the beginning of the modern age of science by Huygens [27], is widely encountered in various fields of science, often observed in living nature [25, 24] and finds a lot of engineering applications [7, 8]. In the classical sense, synchronization means adjustment of frequencies of self-sustained oscillators due to a weak interaction [49].

The history of synchronization goes back to the 17th century when the famous Dutch scientist Christiaan Huygens [27] reported on his observation of synchronization of two pendulum clocks. Systematic study of this phenomenon, experimental as well as theoretical, was started by Edward Appleton [3] and Balthasar van der Pol [70]. They showed that the frequency of a triode generator can be entrained, or synchronized, by a

A. Pikovsky and Y. Maistrenko (eds.), Synchronization: Theory and Application, 187–219.

weak external signal with slightly different frequency. These studies were of high practical importance because such generators became basic elements of radio communication systems.

Next impact to the development of the theory of synchronization was given by the representatives of the Russian school. Andronov and Vitt [2, 1] further developed methods of van der Pol and generalized his results. The case of $n : m$ external synchronization was studied by Mandelshtam and Papaleksi [40]. Mutual synchronization of two weakly nonlinear oscillators was analytically treated by Mayer [41] and Gaponov [23]; relaxation oscillators were studied by Bremsen and Feinberg [9] and Teodorchik [68]. An important step was done by Stratonovich [64, 65] who developed a theory of external synchronization of a weakly nonlinear oscillator in the presence of random noise.

Development of rigorous mathematical tools of the synchronization theory started with Denjoy works on circle map [19] and with treatment of forced relaxation oscillators by Cartwright and Littlewood [11, 12]. Recent development has been highly influenced by Arnold [4] and by Kuramoto [34].

In the context of interacting *chaotic* oscillators, several effects are usually referred to as "synchronization" [49, 47, 35]. Due to a strong interaction of two (or a large number) of identical chaotic systems, their states can coincide, while the dynamics in time remains chaotic [22, 51]. This effect is called "complete synchronization" of chaotic oscillators. It can be generalized to the case of non-identical systems [51, 38, 39], or that of the interacting subsystems [46, 31]. Another well-studied effect is the "chaos–destroying" synchronization, when a periodic external force acting on a chaotic system destroys chaos and a periodic regime appears [36], or, in the case of an irregular forcing, the driven system follows the behavior of the force [32]. This effect occurs for a relatively strong forcing as well. A characteristic feature of these phenomena is the existence of a threshold coupling value depending on the Lyapunov exponents of individual systems [22, 51, 6, 20].

In this article we concentrate on the recently described effect of *phase synchronization* of chaotic systems, which generalizes the classical notion of phase locking. Indeed, for periodic oscillators only the relation between phases is important, while no restriction on the amplitudes is imposed. Thus, we define phase synchronization of chaotic system as an appearance of a certain relation between the phases of interacting systems or between the phase of a system and that of an external force, while the amplitudes can remain chaotic and are, in general, non-correlated. This type of synchronization has been observed in experiments with electronic chaotic oscillators [52, 45], plasma discharge [69], and electrochemical oscillators [28].

2. Synchronization of periodic oscillators

2.1. PHASE LOCKING

In this section we remind basic facts on the synchronization of periodic oscillations (see, e.g., [43]). Stable periodic oscillations are represented by a stable limit cycle in the phase space. The motion of the phase point along the cycle can be parametrized by the phase $\phi(t)$, it's dynamics obeys

$$\frac{d\phi}{dt} = \omega_0, \tag{1}$$

where $\omega_0 = 2\pi/T_0$, and T_0 is the period of the oscillation. It is important that starting from any monotonically growing variable θ on the limit cycle (so that at one rotation θ increases by Θ), one can introduce the phase satisfying Eq. (1). Indeed, an arbitrary θ obeys $\dot{\theta} = \gamma(\theta)$ with a periodic "instantaneous frequency" $\gamma(\theta + \Theta) = \gamma(\theta)$. The change of variables $\phi = \omega_0 \int_0^\theta [\gamma(\theta)]^{-1} d\theta$ gives the correct phase, with the frequency ω_0 being defined from the condition $2\pi = \omega_0 \int_0^\Theta [\nu(\theta)]^{-1} d\theta$. A similar approach leads to correct angle-action variables in Hamiltonian mechanics. We have performed this simple consideration to underline the fact that the notions of the phase and of the phase synchronization are universally applicable to any self-sustained periodic behavior independently on the form of the limit cycle.

From (1) it is evident that the phase corresponds to the zero Lyapunov exponent, while negative exponents correspond to the amplitude variables. Note that we do not consider the equations for the amplitudes, as they are not universal.

When a small external periodic force with frequency ν is acting on this periodic oscillator, the amplitude is relatively robust, so that in the first approximation one can neglect variations of the amplitude to obtain for the phase of the oscillator ϕ and the phase of the external force ψ the equations

$$\frac{d\phi}{dt} = \omega_0 + \varepsilon G(\phi, \psi), \qquad \frac{d\psi}{dt} = \nu, \tag{2}$$

where $G(\cdot, \cdot)$ is 2π-periodic in both arguments and ε measures the strength of the forcing. For a general method of derivation of Eq. (2) see [34]. The system (2) describes a motion on a 2-dimensional torus that appears from the limit cycle under periodic perturbation (see Fig. 1a,b). If we pick up the phase of oscillations ϕ stroboscopically at times $t_n = n\frac{2\pi}{\nu}$, we get a circle map

$$\phi_{n+1} = \phi_n + \varepsilon g(\phi_n), \tag{3}$$

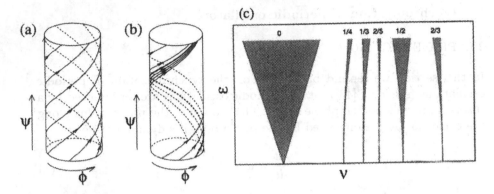

Figure 1. Quasiperiodic (a) and periodic flow (b) on the torus; a stable periodic orbit is shown by the bold line. (c): The typical picture of Arnold tongues (with winding numbers atop) for the circle map.

where the 2π-periodic function $g(\phi)$ is defined via the solutions of the system (2). According to the theory of circle maps (cf. [43]), the dynamics can be characterized by the winding (rotation) number

$$\rho = \lim_{n \to \infty} \frac{\phi_n - \phi_0}{2\pi n},$$

which is independent on the initial point ϕ_0 and can take rational and irrational values. If it is irrational, then the motion is quasiperiodic and the trajectories are dense on the circle. Otherwise, if $\rho = p/q$, there exists a stable orbit with period q such that $\phi_q = \phi_0 + 2\pi p$. The latter regime is called *phase locking* or *synchronization*. In terms of the continuous-time system (2), the winding number is the ratio between the mean derivative of the phase ϕ and the forcing frequency ν

$$\rho = \frac{\left\langle \frac{d\phi}{dt} \right\rangle}{\nu} = \frac{\omega}{\nu}. \tag{4}$$

For ρ irrational and rational one has, respectively, a quasiperiodic dense orbit and a resonant stable periodic orbit on the torus (Fig. 1a,b).

The main synchronization region where $\omega = \nu$ corresponds to the winding number 1 (or, equivalently, 0 if we apply (mod 2π) operation to the phase; for frequencies this means that we consider the difference $\omega - \nu$), other synchronization regions are usually much more narrow. A typical picture of synchronization regions, called also "Arnold tongues", for the circle map (3) is shown in Fig. 1c.

The concept of phase synchronization can be applied only to *autonomous continuous-time* systems. Indeed, if the system is discrete (i.e. a mapping), its period is an integer, and this integer cannot be adjusted to some other integer in a continuous way. The same is true for forced continuous-time oscillations (e.g., for the forced Duffing oscillator): here the frequency of oscillations is completely determined by that of the forcing and cannot be adjusted to some other value. We can formulate this also as follows: in discrete or forced systems there is no zero Lyapunov exponent, so there is no corresponding marginally stable variable (the phase) that can be affected by small external perturbations.

The synchronization condition (4) does not mean that the difference between the phase ϕ of an oscillator and that of the external force ψ (or between phases of two oscillators) must be a constant, as is sometimes assumed (see, e.g. [66]). Indeed, (2) implies, that to enable $\phi - \psi = const$, the function G should depend not on separate phases but only on their difference: $G(\phi, \psi) = G(\phi - \psi)$. Denoting this phase difference as $\varphi = \phi - \psi$ we can rewrite Eq. (2) as

$$\frac{d\varphi}{dt} = \omega_0 - \nu + \varepsilon q(\varphi) \, . \tag{5}$$

In the synchronous state this equation should have (at least one) stable point. This happens if the frequency mismatch (detuning) is small enough, $\varepsilon q_{min} < \nu - \omega_0 < \varepsilon q_{max}$, and this condition determines the synchronization (phase-locking, mode-locking) region on the (ω, ε) plane. Within this region, the phase difference remains constant, $\psi = \delta$, and the value of this constant depends on the detuning, $\delta = q^{-1}[(\nu - \omega_0)/\varepsilon]$ (here the stable branch of the inverse function should be chosen). Generally, the coupling function $G(\phi, \psi)$ cannot be reduced to a function of the phase difference φ. Then, even in a synchronous regime φ is not constant but fluctuates, although these fluctuations are bounded. Thus, we can define phase locking according to relation

$$|\phi(t) - \psi(t) - \delta| < const \, , \tag{6}$$

from which the condition of frequency locking $\langle \dot\phi \rangle = \nu$ naturally follows. The latter definition of phase locking will be used in the treatment of chaotic oscillations below, but even for periodic regimes it has an advantage when the forced oscillations are not close to the original limit cycle.

The winding number is a continuous function of system parameters; typically it looks like a devil's staircase. Take the main phase-locking region. Continuity means that near the de-synchronization transition the mean oscillation frequency is close to the external one. As the external frequency ν is varied, the de-synchronization transition appears as saddle-node bifurcation, where a stable p/q - periodic orbit collides with the corresponding unstable one, and both disappear. Near this bifurcation point, similarly to the type-I intermittency [5], a trajectory of the system spends a large time

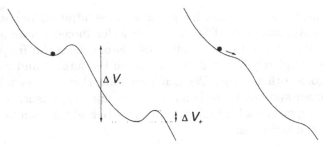

Figure 2.   Phase as a particle in an inclined potential, inside and outside of the synchronization region.

in the vicinity of the just disappeared periodic orbits; in the course of time evolution the long epochs when the phases are locked according to (6), are interrupted with relatively short time intervals where a phase slip (at which the phase difference gains 2π) occurs.

2.2. EFFECT OF NOISE

The most simple way to model a noisy environment is to add a noisy term to the first of Eqs. (2), or, to have the simplest possible situation, to Eq. (5):

$$\frac{d\varphi}{dt} = \omega_0 - \nu + \varepsilon q(\varphi) + \xi(t) . \tag{7}$$

The dynamics of the phase can be treated as the dynamics of an over-damped particle in a potential

$$V(\varphi) = (\nu - \omega_0)\varphi - \varepsilon \int^{\varphi} q(x) \, dx .$$

The average slope of the potential is determined by the mismatch of frequencies of the autonomous oscillator and external force; the depth of the minima (if they exist) is determined by the amplitude of the forcing, see Fig. 2. Without noise, the particle would either rest in a minimum, or slide downwards along the potential, if there are no local minima; this corresponds to a synchronous and non-synchronous states, respectively.

Suppose first that the noise is small and bounded, then its influence results in fluctuations of the particle around a stable equilibrium, i.e. in fluctuations of the phase difference around some constant value. We thus have a situation of phase locking in the sense of relation (6); here the observed frequency coincides with that of the external force.

Contrary to this, if the noise is unbounded (e.g., Gaussian), there is always a probability for the particle to overcome a potential barrier ΔV and to hop in a neighboring minimum of the potential. The time series looks

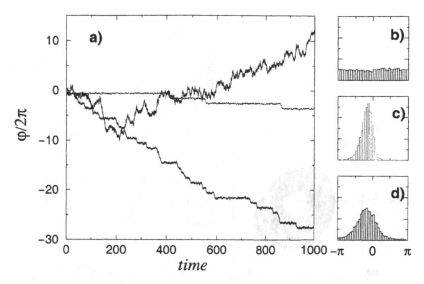

Figure 3. Fluctuation of the phase difference in a noisy oscillator (a). Without forcing, the behavior of φ is diffusive: it performs a motion that reminds a random walk (upper curve); the distribution of φ mod 2π is shown in (b), it is practically uniform. External forcing with non-zero detuning suppresses the diffusion, the phase of the oscillator is nearly locked (middle curve), but sometimes phase slips occur; the respective distribution (c) becomes rather narrow and unimodal. Stronger noise (bottom curve) causes more phase slips, so that there are only rather short epochs where φ oscillates around a constant level; the distribution of φ mod 2π remains nevertheless unimodal (d).

as a sequence of these phase slips (see Fig. 3) and relation (6) does not hold. Nevertheless, at least for small noise, the phase synchronization is definitely detectable, although it is not perfect: between slips we observe epochs of phase locking. Averaged locally over such an epoch, the frequency of the oscillator coincides with that of the external force. The observed frequency that is computed via averaging over a large period of time differs from that of the external force, but this difference is small if the slips are rare.

Phase locking in noisy systems can be also understood in a statistical sense, as an existence of a preferred value of the phase difference φ mod 2π. Indeed, the particle spends most of the time around a position of stable equilibrium, then rather quickly it jumps to a neighboring equilibrium, where the phase difference differs by a multiple of 2π. This can be reflected by distribution of φ mod 2π: a non-synchronous state would have a broad distribution, whereas synchronization would correspond to a unimodal distribution (Fig. 3).

The synchronization transition in noisy oscillators appears as a continuous decrease of characteristic time intervals between slips, and is smeared:

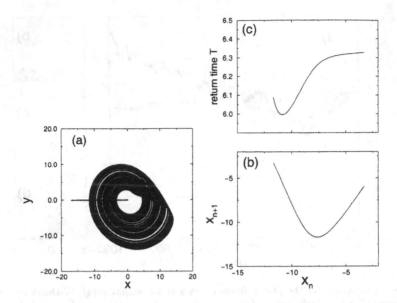

Figure 4. Projection of the phase potrait of the Rössler system (a). The horizontal line shows the Poincaré section that is used for computation of the amplitude mapping (b) and dependence of the return time (rotation period) on the amplitude (c).

we cannot unambiguously determine the border of this transition.

3. Phase and frequency of a chaotic oscillator

3.1. DEFINITION OF THE PHASE

The first problem in extending the basic notions from periodic to chaotic oscillations is to define properly a phase. There seems to be no unambiguous and general definition of phase applicable to an arbitrary chaotic process. Roughly speaking, we want to define phase as a variable which is related to the zero Lyapunov exponent of a continuous-time dynamical system with chaotic behavior. Moreover, we want this phase to correspond to the phase of periodic oscillations satisfying (1).

To be not too abstract, we illustrate a general approach below on the well-known Rössler system. A projection of the phase portrait of this autonomous 3-dimensional system of ODEs (see Eqs. (17) below) is shown in Fig. 4.

Suppose we can define a Poincaré map for our autonomous continuous-time system. Then, for each piece of a trajectory between two cross-sections with the Poincaré surface we define the phase just proportional to time, so that the phase increment is 2π at each rotation:

$$\phi(t) = 2\pi \frac{t - t_n}{t_{n+1} - t_n} + 2\pi n, \qquad t_n \leq t < t_{n+1}. \qquad (8)$$

Here t_n is the time of the n-th crossing of the secant surface. Note that for periodic oscillations corresponding to a fixed point of the Poincaré map, this definition gives the correct phase satisfying Eq. (1). For periodic orbits having many rotations (i.e. corresponding to periodic points of the map) we get a piecewise-linear function of time, moreover, the phase grows by a multiple of 2π during the period. The second property is in fact useful, as it represents in a proper way the organization of periodic orbits inside the chaos. The first property demonstrates that the phase of a chaotic system cannot be defined as unambiguously as for periodic oscillations. In particular, the phase crucially depends on the choice of the Poincaré surface.

Nevertheless, defined in this way, the phase has a physically important property: its perturbations neither grow nor decay in time, so it does correspond to the direction with the zero Lyapunov exponent in the phase space. We note also, that this definition of the phase directly corresponds to the special flow construction which is used in the ergodic theory to describe autonomous continuous-time systems [15].

For the Rössler system (Fig. 4(a)) a proper choice of the Poincaré surface may be the halfplane $y = 0$, $x < 0$. For the amplitude mapping $x_n \to x_{n+1}$ we get a unimodal map (Fig. 4(b), the map is essentially one-dimensional, because the coordinate z for the Rössler attractor is nearly constant on the chosen Poincaré surface). In this and in some other cases the phase portrait looks like rotations around a point that can be taken as the origin, so we can also estimate the phase as the angle between the projection of the phase point on the plane and a given direction on the plane (see also [48, 26]):

$$\phi_P = \arctan(y/x). \qquad (9)$$

Note that although the phase ϕ and its estimate ϕ_P do not coincide microscopically, i.e on a time scale less than the average period of oscillation, they have equal average growth rates. In other words, the mean frequency defined as the average of $d\phi_P/dt$ over large period of time coincides with a straightforward definition of the mean frequency via the average number of crossings of the Poincaré surface per unit time.

3.2. LOCKING-BASED DEFINITION OF FREQUENCY FOR SYSTEMS WITH ILL-DEFINED PHASE

Phase can be straightforwardly introduced if one can find a two-dimensional projection of the attractor in which all trajectories revolve around some origin. This is typically the case for systems exhibiting a transition to chaos via a cascade of period doubling bifurcations; e.g., for the Rössler oscillator. For such projections one can define phase according to Eq. (8), or estimate it according to Eq. (9) or using the Hilbert transform (see section 8 below). Sometimes, a proper projection can be achieved with a coordinate transformation (e.g., using the symmetry properties of the attractor, as in the Lorenz system) [50, 49]. Estimation of the average frequency of individual oscillators $\langle \dot{\phi} \rangle$ then allows one to characterize the degree of synchronization. Contrary to these cases of well-defined phase, chaotic oscillators with "wild", non-revolving trajectories are often termed as those with ill-defined phase. Here only indirect indications for phase synchronization exist (based, e.g., on the ensemble averages [48, 50]), but no direct calculation of the phase and the frequency can be performed.

Nere we describe a method [59], based on the use of *auxiliary limit cycle oscillators*, that allows one to estimate the average frequency of the observed signals for situations with ill-defined phase. To introduce the method let us consider an ensemble of *uncoupled* limit cycle oscillators with natural frequencies ω_k distributed in an interval $[\omega_{min}, \omega_{max}]$. Let each oscillator of this ensemble be driven by a common periodic force of a frequency $\nu \in [\omega_{min}, \omega_{max}]$. It is well-known that the force synchronizes those elements of the ensemble which have frequencies close to ν. This can be demonstrated by plotting the frequencies of the driven limit cycle oscillators Ω_k, called hereafter the observed frequencies, vs. the natural frequencies ω_k: the synchronization manifests itself in the appearance of a horizontal plateau (more precisely, one expects to observe a devil's staircase structure with infinitely many plateaus), where the frequency of entrained elements is equal to ν. Hence, an *unknown* frequency of the drive can be revealed by the analysis of the Ω_k vs. ω_k plot. The idea of our approach is to use the ensemble of auxiliary oscillators as a *device for measuring the frequency of complex signals*.

A simple implementation of the method is to drive the array of Poincaré oscillators with a signal $X(t)$, which frequency we would like to determine

$$\dot{A}_k = (1 + i\omega_k)A_k - |A_k|^2 A_k + \varepsilon X(t) . \tag{10}$$

Separating the real amplitude R and the phase ϕ from the complex amplitude $A = Re^{i\phi}$ we obtain for the phase $\dot{\phi}_k = \omega_k - R_k^{-1}\varepsilon X(t)\sin\phi_k$. Noting that for small ε the amplitude R is close to unity, and neglecting its fluctuations, we can write equations for our measuring oscillators as pure phase equations:

$$\dot{\phi}_k = \omega_k - \varepsilon X(t)\sin\phi_k, \tag{11}$$

and the observed frequencies are $\Omega_k = \langle \dot{\phi}_k \rangle$. Note that Eqs. (11) become exact if one writes a higher order nonlinearity $|A|^p A$ in (10) and considers the limit $p \to \infty$. In calculations below we normalize the signal $X(t)$ to have zero mean and unit variance so that the coupling constant ε is the only parameter of the method (the mean value can slightly influence the result).

To show how the method works we consider a model quasiharmonic process with mean frequency ω_0 and slowly varying amplitude and phase: $X(t) = 2(1 + a(t)) \cos(\omega_0 t - \theta(t))$. Substituting this in (11) and averaging over the period of fast oscillations $2\pi/\omega_0$, we obtain for the slowly varying phase difference $\psi = \phi - \omega_0 t + \theta$ the equation

$$\dot{\psi} = \omega - \omega_0 + \dot{\theta}(t) - \varepsilon(1 + a(t)) \sin \psi \, ,$$

for a harmonic signal ($\dot{\theta} = a = 0$) it has for $\varepsilon \geq |\omega - \omega_0|$ the synchronized solution $\psi_0 = \arcsin((\omega - \omega_0)/\varepsilon)$. For weak modulation we can linearize around this state and obtain for the deviations $\delta\psi$:

$$\frac{d(\delta\psi)}{dt} = \dot{\theta} - a(t)(\omega - \omega_0) - \sqrt{\varepsilon^2 - (\omega - \omega_0)^2} \delta\psi \, .$$

Assuming that θ and a are independent random processes, we can express the power spectrum of the phase fluctuations through the spectra of these processes:

$$S_{\delta\psi}(\sigma) = \frac{\sigma^2 S_\theta(\sigma) + (\omega - \omega_0)^2 S_a(\sigma)}{\varepsilon^2 - (\omega - \omega_0)^2 + \sigma^2} \, .$$

One can see that the fluctuations are small only in the middle of the synchronization region (for $\omega \approx \omega_0$; here only the phase fluctuations S_θ contribute. Modeling S_θ by the Lorentz-like spectrum

$$S_\theta = \frac{2\Delta V_\theta}{(\sigma^2 + \Delta^2)\pi} \, ,$$

where V_θ and Δ are the variance and the characteristic maximal frequency of fluctuations of θ, we obtain

$$V_{\delta\psi} = \int_0^\infty S_{\delta\psi}(\sigma) d\sigma = V_\theta \Delta(\varepsilon + \Delta)^{-1} \, .$$

This final formula shows that good synchronization (i.e. small variance of $\delta\psi$) can be achieved if ε is sufficiently larger than Δ, i.e. if the coupling constant is larger than the characteristic frequency of phase fluctuations. From the other side, in the limit $\varepsilon \to \infty$ the dependence of the observed frequency Ω on the natural one disappears, and the measured frequency is the Rice frequency of the process $X(t)$.

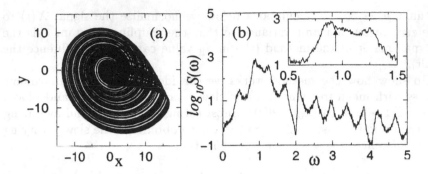

Figure 5. (a) Funnel attractor in the Rössler system (12). (b) Power spectrum of $x(t)$. The arrow shows the characteristic frequency as determined according to Fig. 6.

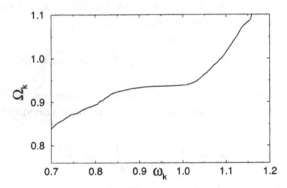

Figure 6. Output of the frequency measuring "device" (11) as a function of the natural frequencies ω_k for the Rössler oscillator (12). The height of the plateau determines the characteristic frequency of this chaotic drive $\Omega^{(p)} = 0.94$.

To illustrate the approach we consider the Rössler system with a funnel attractor (Fig. 5a):

$$
\begin{aligned}
\dot{x} &= -y - z \,, \\
\dot{y} &= x + 0.4y \,, \\
\dot{z} &= 0.25 + z(x - 8.5) \,.
\end{aligned}
\tag{12}
$$

Clearly we cannot find an origin around which all trajectories revolve. The power spectrum for the variable x (Fig. 5b) is broad; it does not contain a dominating maximum. Due to these properties, there is no direct way to introduce the phase for this system and to characterize its synchronization [50]. In our method, we use the (normalized) signal $x(t)$ to drive system (11) with $\varepsilon = 0.5$. The frequencies of the oscillators in the measuring device driven by $X(t) = x(t)$ are shown in Fig. 6. The resulting plateau in the Ω_k vs. ω_k plot gives $\Omega^p \approx 0.94$ where "p" stands for "plateau". Practically, the middle point of the plateau Ω_k^p was determined from the minimum of the running variance $\sum_{j=k-L}^{k+L}(\Omega_j - \bar{\Omega}_k)^2$, where $\bar{\Omega}_k = (2L+1)^{-1}\sum_{j=k-L}^{k+L}\Omega_j$. Variation of L from 3 to 10 gave no essential difference. This method provides also a smoothening of the Ω_k vs. ω_k curve. One can see that this characteristic frequency cannot be directly associated with a peak in the power spectrum (Fig. 5b). We also see that Fig. 6 does not show the devil's staircase structure, but only one, smeared plateau. This is due to the chaotic nature of the process $x(t)$, so that, similar to the case of narrow-band noisy signals, the high-order phase-locked regions are not observed [37, 49].

3.3. DYNAMICS OF THE PHASE OF CHAOTIC OSCILLATIONS

In contrast to the dynamics of the phase of periodic oscillations, the growth of the phase in the chaotic case cannot generally be expected to be uniform. Instead, the instantaneous frequency depends in general on the amplitude. Let us hold to the phase definition based on the Poincaré map, so one can represent the dynamics as (cf. [52])

$$A_{n+1} = M(A_n),\qquad(13)$$

$$\frac{d\phi}{dt} = \omega(A_n) \equiv \omega_0 + F(A_n).\qquad(14)$$

As the amplitude A we take the set of coordinates for the point on the secant surface; it does not change during the growth of the phase from 0 to 2π and can be considered as a discrete variable; the transformation M defines the Poincaré map. The phase evolves according to (14), where the "instantaneous" frequency $\omega = 2\pi/(t_{n+1} - t_n)$ depends in general on the amplitude. Assuming the chaotic behavior of the amplitudes, we can consider the term $\omega(A_n)$ as a sum of the averaged frequency ω_0 and of some effective noise $F(A)$; in exceptional cases $F(A)$ may vanish. For the Rössler attractor the "period" of the rotations (i.e. the function $2\pi/\omega(A_n)$) is shown in Fig. 4(c). This period is not constant, so the function $F(A)$ does not vanish, but the variations of the period are relatively small.

Hence, the Eq. (14) is similar to the equation describing the evolution of phase of periodic oscillator in the presence of external noise. Thus, the dynamics of the phase is generally diffusive: for large t one expects

$$\langle(\phi(t) - \phi(0) - \omega_0 t)^2\rangle \propto D_p t,$$

where the diffusion constant D_p determines the phase coherence of the chaotic oscillations. Roughly speaking, the diffusion constant is proportional to the width of the spectral peak calculated for the chaotic observable [21].

Generalizing Eq. (14) in the spirit of the theory of periodic oscillations to the case of periodic external force, we can write for the phase

$$\frac{d\phi}{dt} = \omega_0 + \varepsilon G(\phi, \psi) + F(A_n), \qquad \frac{d\psi}{dt} = \nu. \tag{15}$$

Here we assume that the force is small (of order of ε) so that it affects only the phase, and the amplitude obeys therefore the unperturbed mapping M. This equation is similar to Eq. (7), with the amplitude-depending part of the instantaneous frequency playing the role of noise. Thus, we expect that in general the synchronization phenomena for periodically forced chaotic system are similar to those in noisy driven periodic oscillations. One should be aware, however, that the "noisy" term $F(A)$ can be hardly calculated explicitly, and for sure cannot be considered as a Gaussian δ-correlated noise as is commonly assumed in the statistical approaches [65, 55].

4. Phase synchronization by external force

4.1. SYNCHRONIZATION REGION

We describe here the effect of phase synchronization of chaotic oscillations by periodic external force, taking as examples two prototypic models of nonlinear dynamics: the Lorenz

$$\begin{aligned}
\dot{x} &= 10(y - x), \\
\dot{y} &= 28x - y - xz, \\
\dot{z} &= -8/3 \cdot z + xy + E\cos\nu t.
\end{aligned} \tag{16}$$

and the Rössler

$$\begin{aligned}
\dot{x} &= -y - z + E\cos\nu t, \\
\dot{y} &= x + 0.15y, \\
\dot{z} &= 0.4 + z(x - 8.5).
\end{aligned} \tag{17}$$

oscillators. In the absence of forcing, both are 3-dimensional dissipative systems which admit a straightforward construction of the Poincaré maps. Moreover, we can simply use the phase definition (9), taking the original variables (x, y) for the Rössler system and the variables $(\sqrt{x^2 + y^2} - u_0, z - z_0)$ for the Lorenz system (where $u_0 = 12\sqrt{2}$ and $z_0 = 27$ are the coordinates of the equilibrium point, the "center of rotation"). The mean rotation frequency can be thus calculated as

$$\Omega = \lim_{t \to \infty} 2\pi \frac{N_t}{t} \tag{18}$$

where N_t is the number of crossings of the Poincaré section during observation time t. This method can be straightforwardly applied to the observed

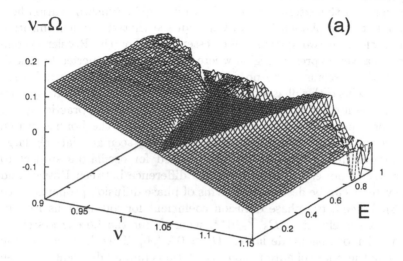

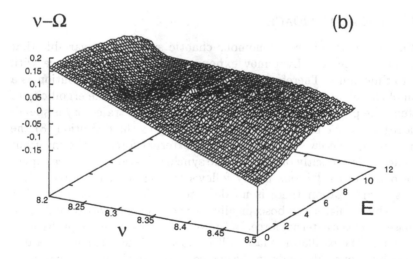

Figure 7. The phase synchronization regions for the Rössler (a) and the Lorenz (b) systems.

time series, in the simplest case one can, e.g., take for N_t the number of maxima (of $x(t)$ for the Rössler system and of $z(t)$ for the Lorenz one).

Dependence of the frequency Ω obtained in this way on the amplitude and frequency of the external force is shown in Fig. 7. Synchronization here corresponds to the plateau $\Omega = \nu$. One can see that the synchronization properties of these two systems differ essentially. For the Rössler system there exists a well-expressed region where the systems are perfectly locked. Moreover, there seems to be no amplitude threshold of synchronization (cf. Fig. 1c, where the phase-locking regions start at $\varepsilon = 0$). It appears that the phase locking properties of the Rössler system are practically the same as for a periodic oscillator. On the contrary, for the Lorenz system we observe the frequency locking only as a tendency seen at relatively large forcing amplitudes, as this should be expected for oscillators subject to a rather strong noise. In this respect, the difference between Rössler and Lorenz systems can be described in terms of phase diffusion properties (see Sect. 3.3). Indeed, the phase diffusion coefficient for autonomous Rössler system is extremely small $D_p < 10^{-4}$, whereas for the Lorenz system it is several oder of magnitude larger, $D_p \approx 0.2$ [50]. This difference in the coherence of the phase of autonomous oscillations implies different response to periodic forcing.

In the following sections we discuss the phase synchronization of chaotic oscillations from the statistical and the topological viewpoints.

4.2. STATISTICAL APPROACH

We define the phase of an autonomous chaotic system as a variable that corresponds to the zero Lyapunov exponent, i.e. to the invariance with respect to time shifts. Therefore, the invariant probability distribution as a function of the phase is nearly uniform. This follows from the ergodicity of the system: the probability is proportional to the time a trajectory is spending in a region of the phase space, and according to the definition (8) the phase motion is (piecewise) uniform. With external forcing, the invariant measure depends explicitly on time. In the synchronization region we expect that the phase of oscillations nearly follows the phase of the force, while without synchronization there is no definite relation between them. Let us observe the oscillator stroboscopically, at the moments corresponding to some phase ψ_0 of the external force. In the synchronous state the probability distribution of the oscillator phase will be localized near some preferable value (which of course depends on the choice of ψ_0). In the non-synchronous state the phase is spread along the attractor. We illustrate this behavior of the probability density in Fig. 8. One can say that synchronization means localization of the probability density near some preferable time-periodic state. In other words, this means appearance of the long-range correlation

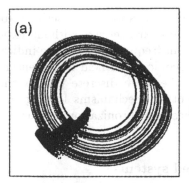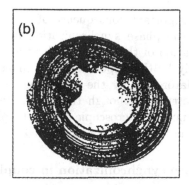

Figure 8. Distribution inside (a) and outside (b) the synchronization region for the Rössler system, shown with black dots. The autonomous Rössler attractor is shown with gray.

in time and of the significant discrete component in the power spectrum of oscillations.

Let us consider now the ensemble interpretation of the probability. Suppose we take a large ensemble of identical copies of the chaotic oscillator which differ only by their initial states, and let them evolve under the same periodic forcing. After the transient, the projections of the phase state of each oscillator onto the plane x, y form the cloud that exactly corresponds to the probability density. Let us now consider the ensemble average of some observable. Without synchronization the cloud is spread over the projection of the attractor (Fig. 8b), and the average is small: no significant average field is observed. In the synchronous state the probability is localized (Fig. 8a), so the average is close to some middle point of the cloud; this point rotates with the frequency ν and one observes large regular oscillations of the average field. Hence, the synchronization can be easily indicated through the appearance of a large (macroscopic) mean field in the ensemble. Physically, this effect is rather clear: unforced chaotic oscillators are not coherent due to internal chaos, thus the summation of their fields yields a small quantity. Being synchronized, the oscillators become coherent with the external force and thereby with each other, so the coherent summation of their fields produces a large mean field.

We can look on the probability also from the ergodic point of view, where instead of taking an ensemble one takes separated in time states of one system. The described above coherence that appears due to phase synchronization is now coherence in time. It can be revealed by calculating the autocorrelation function or the power spectrum. Synchronization means that correlations in time are large and a significant discrete peak appears in the spectrum of oscillations.

An important consequence of the statistical approach described above is that the phase synchronization can be characterized without explicit computation of the phase and/or the mean frequency: it can be indicated implicitly by the appearance of a macroscopic mean field in the ensemble of oscillators, or by the appearance of the large discrete component in the spectrum. Although there may be other mechanisms leading to the appearance of macroscopic order, the phase synchronization appears to be one of the most common ones.

5. Phase synchronization in coupled systems

Now we demonstrate the effects of phase synchronization in coupled chaotic oscillators. We start with the simplest case of two interacting systems, and then briefly discuss oscillator lattices, globally coupled systems, and space-time chaos.

5.1. SYNCHRONIZATION OF TWO INTERACTING OSCILLATORS

We consider here two non-identical coupled Rössler systems

$$
\begin{aligned}
\dot{x}_{1,2} &= -\omega_{1,2}y_{1,2} - z_{1,2} + \varepsilon(x_{2,1} - x_{1,2}), \\
\dot{y}_{1,2} &= \omega_{1,2}x_{1,2} + ay_{1,2}, \\
\dot{z}_{1,2} &= f + z_{1,2}(x_{1,2} - c),
\end{aligned}
\tag{19}
$$

where $a = 0.165$, $f = 0.2$, $c = 10$. The parameters $\omega_{1,2} = \omega_0 \pm \Delta\omega$ and ε determine the mismatch of natural frequencies and the coupling, respectively.

Again, like in the case of periodic forcing, we can define the mean frequencies $\Omega_{1,2}$ of oscillations of each system, and study the dependence of the frequency mismatch $\Omega_2 - \Omega_1$ on the parameters $\Delta\omega, \varepsilon$. This dependence is shown in Fig. 9 and demonstrates a large region of synchronization between two oscillators.

It is instructive to characterize the synchronization transition by means of the Lyapunov exponents (LE). The 6-order dynamical system (19) has 6 LEs (see Fig. 10). For zero coupling we have a degenerate situation of two independent systems, each of them has one positive, one zero, and one negative exponent. The two zero exponents correspond to the two independent phases. With coupling, the phases become dependent and the degeneracy must be removed: only one LE should remain exactly zero. We observe, however, that for small coupling also the second zero Lyapunov exponent remains extremely small (in fact, numerically indistinguishable from zero). Only at relatively stronger coupling, when the synchronization sets on, the second LE becomes negative: now the phases are dependent

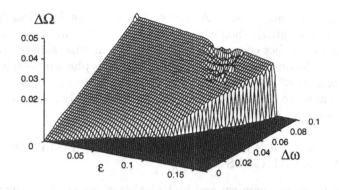

Figure 9. Synchronization of two coupled Rössler oscillators; $\omega_0 = 1$.

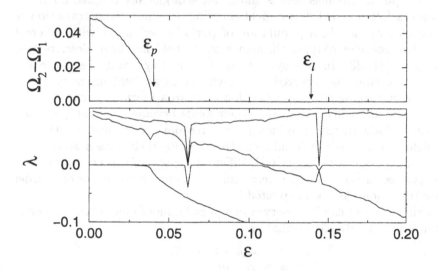

Figure 10. The Lyapunov exponents λ (bottom panel, only the 4 largest LEs are depicted) and the frequency difference vs. the coupling ε in the coupled Rössler oscillators; $\omega_0 = 0.97$, $\Delta\omega = 0.02$. Transition to the phase (ε_p) and to the lag synchronization (ε_l) are marked.

and a relation between them is stable. Note that the two positive exponents remain positive which means that the amplitudes remain chaotic and independent: the coupled system remains in the state of hyperchaos.

With the increase of coupling one of the positive LE becomes smaller. Physically this means that not only the phases are locked, but the difference between the amplitudes is suppressed by coupling as well. At a certain coupling only one LE remains positive, so one can expect synchronization both in phases and amplitudes. As the systems are not identical (due to the frequency mismatch), their states cannot be identical: $x_1(t) \neq x_2(t)$. However, almost perfect correspondence between the time-shifted states of the systems can be observed: $x_1(t) \approx x_2(t - \Delta t)$. This phenomenon is called "lag synchronization" [58]. With further increase of the coupling ε the lag Δt decreases and the states of two systems become nearly identical, like in case of complete synchronization.

5.2. SYNCHRONIZATION IN A POPULATION OF GLOBALLY COUPLED CHAOTIC OSCILLATORS

A number of physical, chemical and biological systems can be viewed at as large populations of weakly interacting non-identical oscillators [34]. One of the most popular models here is an ensemble of globally coupled nonlinear oscillators (often called "mean-field coupling"). A nontrivial transition to self-synchronization in a population of periodic oscillators with different natural frequencies coupled through a mean field has been described by Kuramoto [34, 33]. In this system, as the coupling parameter increases, a sharp transition is observed for which the mean field intensity serves as an order parameter. This transition owes to a mutual synchronization of the periodic oscillators, so that their fields become coherent (i.e. their phases are locked), thus producing a macroscopic mean field. In its turn, this field acts on the individual oscillators, locking their phases, so that the synchronous state is self-sustained. Different aspects of this transition have been studied in [61, 17, 18], where also an analogy with the second–order phase transition has been exploited.

A similar effect can be observed in a population of *non-identical chaotic* systems, e.g. the Rössler oscillators

$$\begin{aligned}
\dot{x}_i &= -\omega_i y_i - z_i + \varepsilon X, \\
\dot{y}_i &= \omega_i x_i + a y_i, \\
\dot{z}_i &= 0.4 + z_i(x_i - 8.5),
\end{aligned} \qquad (20)$$

coupled via the mean field $X = N^{-1} \sum_1^N x_i$. Here N is the number of elements in the ensemble, ε is the coupling constant, a and ω_i are parameters of the Rössler oscillators. The parameter ω_i governs the natural frequency of an individual system. We take a set of frequencies ω_i which are Gaussian-distributed around the mean value ω_0 with variance $(\Delta\omega)^2$. The Rössler system typically shows windows of periodic behavior as the parameter ω is changed; therefore we usually choose a mean frequency ω_0 in a way that we avoid large periodic windows. In our computer simulations we solve numerically Eqs. (20) for rather large ensembles $N = 3000 \div 5000$.

With an increase of the coupling strength ε, the appearance of a non-zero macroscopic mean field X is observed [48], as is shown in Fig. 11. This

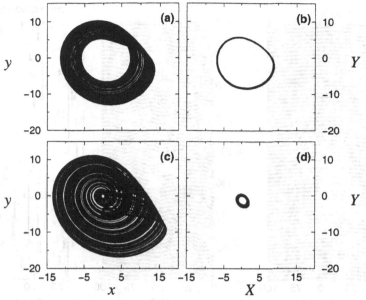

Figure 11.   Projections of the phase portraits of the Rössler oscillators (left column) and of the mean fields $X = \langle x_i \rangle$, $Y = \langle y_i \rangle$ in ensemble of $N = 5000$ oscillators. (a): Phase-coherent Rössler attractor, $\omega_0 = 1$, $a = 0.15$. (b): Mean field in the ensemble of oscillators (a) with Gaussian distribution of frequencies $\Delta\omega = 0.02$ and coupling $\varepsilon = 0.1$. (c) Funnel attractor $\omega_0 = 0.97$, $a = 0.25$. (d): Mean field in the ensemble of oscillators (c) with Gaussian distribution of frequencies $\Delta\omega = 0.02$ and coupling $\varepsilon = 0.15$.

indicates the phase synchronization of the Rössler oscillators that arises due to their interaction via mean field. This mean field is large, if the attractors of individual systems are phase-coherent (parameter $a = 0.15$) and the phase is well-defined. On the contrary, in the case of the funnel attractor $a = 0.25$, when the oscillations look wild and the imaging point makes large and small loops around the origin, there seems to be no way to choose the Poincaré section unambiguously; in this case the mean field is rather small. Nevertheless, in both cases synchronization transition is clearly indicated by the onset of the mean field, without computation of the phases themselves. Finally, we note that phase synchronization in a globally coupled ensemble of chaotic oscillators have been observed experimentally in [30, 29].

6. Lattice of chaotic oscillators

If chaotic oscillators are ordered in space and form a lattice, usually it is assumed that only the nearest neighbors interact. Such a situation is relevant for chemical systems, where homogeneous oscillations are chaotic, and the diffusive coupling can be modeled with dissipative nearest neighbors

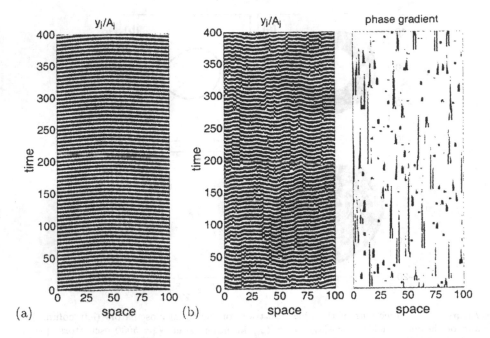

Figure 12.   (a): Nearly homogeneous phase profile for the lattice of identical phase-coherent Rössler oscillators. (b) In a lattice of identical funnel Rössler oscillators $a = 0.23$, $\varepsilon = 0.05$ an inhomogeneous phase profile with numerous defects is observed. The defects are clearly seen in the right panel as large values of the gradient

interaction [10, 26]. In a lattice, one can expect complex spatio-temporal synchronization structures to be observed.

Consider as a model a 1-dimensional lattice of Rössler oscillators with local dissipative coupling:

$$\begin{aligned}
\dot{x}_j &= -\omega_j y_j - z_j, \\
\dot{y}_j &= \omega_j x_j + a y_j + \varepsilon(y_{j+1} - 2y_j + y_{j-1}), \\
\dot{z}_j &= 0.4 + (x_j - 8.5)z_j.
\end{aligned} \qquad (21)$$

Here the index $j = 1, \ldots, N$ counts the oscillators in the lattice and ε is the coupling coefficient. In a homogeneous lattice (i.e. for equal natural frequencies ω_j) the observed regime significantly depends on the coherence properties of a single chaotic oscillator. If the Rössler oscillator is phase-coherent, all the phases nearly synchronize and a regular phase pattern (in fact, a nearly homogeneous phase distribution) is observed (Fig. 12a). In the case of funnel Rössler oscillator the phase at a individual element can "spontaneously" change by π, this prevents synchronization in the lattice (Fig. 12b).

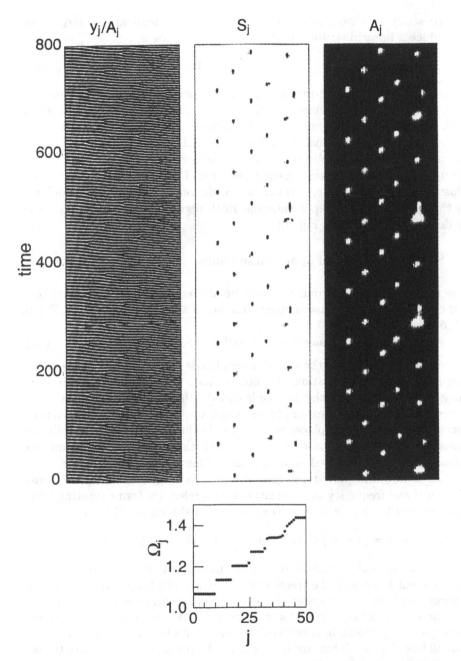

Figure 13.   Clusters and space-time defects in a lattice of Rössler oscillators with a linear distribution of natural frequencies. On the bottom panel the observed frequencies are shown.

To study synchronization in a lattice of non-identical oscillators, we introduce a linear distribution of natural frequencies ω_j

$$\omega_j = \omega_1 + \delta(j - 1) \tag{22}$$

where δ is the frequency mismatch between neighboring sites. Depending on the values of δ we observed two scenarios of transition to synchronization [42]. For small δ, the transition occurs smoothly, i.e. all the elements along the chain gradually adjust their frequencies. If the frequency mismatch is larger, clustering is observed: the oscillators build phase-synchronized groups having different mean frequencies (Fig. 13). At the borders between clusters phase slips occur; this can be considered as appearance of defects in the spatio-temporal representation. Both regular and irregular patterns of defects can be seen in Fig. 13.

7. Synchronization of space-time chaos

The idea of phase synchronization can be also applied to space-time chaos. For example, in the famous complex Ginzburg-Landau equation (CGLE) [16, 13, 63]

$$\partial_t a = (1 + i\omega_0)a - (1 + i\alpha)|a|^2 a + (1 + i\beta)\partial_t^2 a , \tag{23}$$

there are regimes where the complex amplitude a rotates with some mean frequency, but these rotations are not regular: the phase deviates irregularly in space and time (this regime is called "phase turbulence"). Another regime, where the complex amplitude a not always rotates but experiences space-time defects (the places where the absolute value of the amplitude vanishes whereas the phase is not defined); this regime is analogous to oscillations with ill-defined phase like in the funnel Rössler oscillator.

Let us now add periodic in time spatially homogeneous forcing of amplitude B and frequency ω_e. Transition into a reference frame rotating with this external forcing $(a \rightarrow A \equiv a \exp(-i\omega_e t))$ reduces Eq. (23) to

$$\partial_t A = (1 + i\nu)A - (1 + i\alpha)|A|^2 A + (1 + i\beta)\partial_t^2 A + B , \tag{24}$$

where $\nu = \omega_0 - \omega_e$ is the frequency mismatch between the frequency of the external force and the frequency of small oscillations. An analysis of different regimes in the system (24) has been recently performed [14]. As one can expect, a very strong force suppresses turbulence and the spatially homogeneous periodic in time synchronous oscillations are observed, while a small force has no significant influence on the turbulent state. A nontrivial regime is observed for intermediate forcing: in some parameter range the irregular fluctuations of the phase are not completely suppressed but are bounded: the whole system oscillates "in phase" with the external force

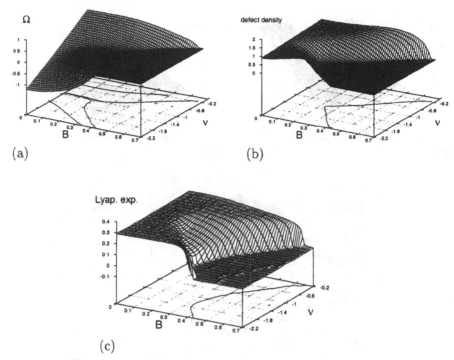

Figure 14. Synchronization of defect turbulence ($\alpha = -2$, $\beta = 2$): (a) Average frequency as a function of the amplitude B and the frequency ν of the external force. The contour lines are drawn at levels -0.5, -0.01, 0.01, 0.5. (b) Density of defects (arbitrary units). The contour line shows the border of the defect-free region. (c) The largest Lyapunov exponent. The contour line shows the border of the turbulent region.

and is highly coherent, although some small chaotic variations persist. In Figs. 14,15 we show how the forcing acts on the regimes with defect and phase turbulence in the CGLE. One can see that the forcing can supress defects while not supressing completely the space-time chaos; this regime is analogous to phase synchronization of individual oscillators. In the case of phase turbulence there is a range of parameters where defects are observed, they appear due to a specific for the forced CGLE instability of special solutions – kinks – connecting two spatially homogeneous regions of complete synchronization, the phases in these regions differ by $\pm 2\pi$. These kinks can disappear through a defect, but from this defect two new kinks appear, leading to kink-breeding process (Fig. 16).

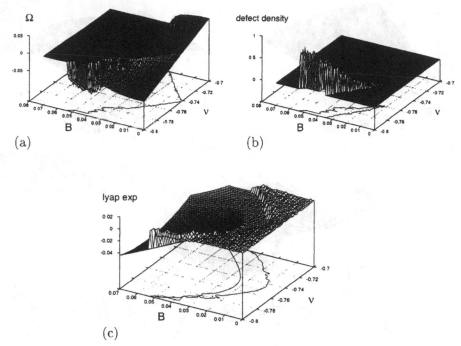

(a)   (b)

(c)

Figure 15.   Synchronization of phase turbulence ($\alpha = -0.75$, $\beta = 2$). (a) The average frequency Ω as a function of the amplitude B and the frequency ν of the external force. The contour line shows the border of the synchronization region. (b) The density of defects (arbitrary units). (c) The largest Lyapunov exponent. The contour lines show the borders between the regions of positive, zero, and negative exponents.

8. Detecting synchronization in data

The analysis of relation between the phases of two systems, naturally arising in the context of synchronization, can be used to approach a general problem in time series analysis. Indeed, bivariate data are often encountered in the study of real systems, and the usual aim of the analysis of such data is to find out whether two signals are dependent or not. As experimental data are very often non-stationary, the traditional techniques, such as cross–spectrum and cross–correlation analysis [44], or non–linear characteristics like generalized mutual information [53] or maximal correlation [71] have their limitations. From the other side, sometimes it is reasonable to assume that the observed signals originate from two weakly interacting systems. The presence of this interaction can be found by means of the analysis of *instantaneous* phases of these signals. These phases can be unambigu-

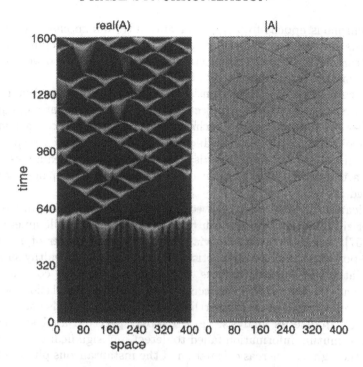

Figure 16.   Kink-breeding process in the phase turbulence for $\alpha = -0.75$, $\beta = 2$, $\nu = -0.8$, and $B = 0.041$. Initial conditions are small fluctuations around the unstable synchronized plane wave. Grey scale space-time plots of $Re(A)$ and $|A|$ reveal the occurrence of defects (black spots in the right panel) corresponding to zeros of $|A|$.

ously obtained with the help of the analytic signal concept based on the Hilbert transform (for an introduction see [44, 50]). It goes as follows: for an arbitrary scalar signal $s(t)$ one can construct a complex function of time (analytic signal) $\zeta(t) = s(t) + i\tilde{s}(t) = A(t)e^{i\phi_H(t)}$ where $\tilde{s}(t)$ is the Hilbert transform of $s(t)$,

$$\tilde{s}(t) = \pi^{-1}\mathrm{P.V.}\int_{-\infty}^{\infty}\frac{s(\tau)}{t - \tau}d\tau \,, \qquad (25)$$

and $A(t)$ and $\phi_H(t)$ are the instantaneous amplitude and phase (P.V. means that the integral is taken in the sense of the Cauchy principal value). Alternatively, one can use a wavelet transform with a complex Gabor wavelet, see [54].

As has been recently shown in [56, 50], the phase estimated by this method from an appropriately chosen oscillatory observable practically coincides with the phase of an oscillator computed according to one of the definitions given in Sec. 3. Therefore, the analysis of the relationship between these Hilbert phases appears to be an appropriate tool to detect synchronous epochs from experimental data and to check for a weak interaction between systems under study. It is very important that the Hilbert transform does not require stationarity of the data, so we can trace synchronization transitions even from nonstationary data.

We recall again the above mentioned similarity of phase dynamics in noisy and chaotic oscillators (see Sect. 3.3). A very important consequence of this fact is that, using the synchronization approach to data analysis, we can avoid the hardly solvable dilemma "noise vs chaos": irrespectively of the origin of the observed signals, the approach and techniques of the analysis are unique. Quantification of synchronization from noisy data is considered in [67].

Application of these ideas allowed us to find phase locking in the data characterizing mechanisms of posture control in humans while quiet standing [60, 57]. Namely, the small deviations of the body center of gravity in anterior–posterior and lateral directions were analyzed. In healthy subjects, the regulation of posture in these two directions can be considered as independent processes, and the occurrence of some interrelation possibly indicates a pathology. It is noteworthy that in several records conventional methods of time series analysis, i.e. the cross–spectrum analysis and the generalized mutual information failed to detect any significant dependence between the signals, whereas calculation of the instantaneous phases clearly showed phase locking.

Complex synchronous patterns have been found recently in the analysis of interaction of human cardiovascular and respiratory systems [62]. This finding possibly indicates the existence of a previously unknown type of neural coupling between these systems.

Analysis of synchronization between brain and muscle activity of a Parkinsonian patient [67] is relevant for a fundamental problem of neuroscience: can one consider the synchronization between different areas of the motor cortex as a necessary condition for establishing of the coordinated muscle activity? It was shown [67] that the temporal evolution of the coordinated pathologic tremor activity directly reflects the evolution of the strength of synchronization within a neural network involving cortical motor areas. Additionally, the brain areas with the tremor-related activity were localized from noninvasive measurements.

9. Conclusions

The main idea of this paper is to demonstrate that synchronization phenomena in periodic, noisy and chaotic oscillators can be understood within a unified framework. This is achieved by extending the notion of phase to the case of continuous-time chaotic systems. Because the phase is introduced as a variable corresponding to the zero Lyapunov exponent, this notion should be applicable to any autonomous chaotic oscillator. Although we are not able to propose a unique and rigorous approach to determine the phase, we have shown that it can be introduced in a reasonable and consistent way for basic models of chaotic dynamics. Moreover, we have shown that even in the case when the phases are not well-defined, i.e. they cannot be unambiguously computed explicitly, the presence of phase synchronization can be demonstrated indirectly by observations of the mean field and the spectrum, i.e. independently of any particular definition of the phase.

In a rather general framework, any type of synchronization can be considered as appearance of some additional order inside the dynamics. For chaotic systems, e.g., the complete synchronization means that the dynamics in the phase space is restricted to a symmetrical submanifold. Thus, from the point of view of topological properties of chaos, the synchronization transition usually means the simplification of the structure of the strange attractor. In discussing the topological properties of phase synchronization, we have shown that the transition to phase synchronization corresponds to splitting of the complex invariant chaotic set into distinctive attractor and repeller. Analogously to the complete synchronization, which appears through the pitchfork bifurcation of the strange attractor, one can say that the phase synchronization appears through tangent bifurcation of strange sets.

Because of the similarity in the phase dynamics, one may expect that many, if not all, synchronization features known for periodic oscillators can be observed for chaotic systems as well. Indeed, here we have described effects of phase and frequency entrainment by periodic external driving, both for simple and space-distributed chaotic systems. Further, we have described synchronization due to interaction of two chaotic oscillators, as well as self-synchronization in globally coupled large ensembles.

As an application of the developed framework we have discussed a problem in data analysis, namely detection of weak interaction between systems from bivariate data. The three described examples of the analysis of physiological data demonstrate a possibility to detect and characterize synchronization even from nonstationary and noisy data.

Finally, we would like to stress that contrary to other types of chaotic synchronization, the phase synchronization phenomena can happen already for very weak coupling, which offers an easy way of chaos regulation.

Acknowledgements

We thank M. Zaks, J. Kurths, G. Osipov, H. Chaté, O. Rudzick, U. Parlitz, P. Tass, C. Schäfer for useful discussions.

References

1. Andronov, A. A. and A. A. Vitt: 1930a, 'On mathematical theory of entrainment'. *Zhurnal prikladnoi fiziki (J. Appl. Phys.)* **7**(4), 3. (In Russian).
2. Andronov, A. A. and A. A. Vitt: 1930b, 'Zur Theorie des Mitnehmens von van der Pol'. *Archiv für Elektrotechnik* **24**(1), 99–110.
3. Appleton, E. V.: 1922, 'The Automatic Synchronization of Triode Oscillator'. *Proc. Cambridge Phil. Soc. (Math. and Phys. Sci.)* **21**, 231–248.
4. Arnold, V. I.: 1961, 'Small denominators. I. Mappings of the circumference onto itself'. *Izv. Akad. Nauk Ser. Mat.* **25**(1), 21–86. (In Russian); English Translation: AMS Transl. Ser. 2, v. 46, 213-284.
5. Bergé, P., Y. Pomeau, and C. Vidal: 1986, *Order within chaos*. New York: Wiley.
6. Bezaeva, L., L. Kaptsov, and P. S. Landa: 1987, 'Synchronization Threshold as the Criterium of Stochasticity in the Generator with Inertial Nonlinearity'. *Zhurnal Tekhnicheskoi Fiziki* **32**, 467–650. (In Russian).
7. Blekhman, I. I.: 1971, *Synchronization of Dynamical Systems*. Moscow: Nauka. (In Russian).
8. Blekhman, I. I.: 1981, *Synchronization in Science and Technology*. Moscow: Nauka. (In Russian); English translation: 1988, ASME Press, New York.
9. Bremsen, A. S. and I. S. Feinberg: 1941, 'Analysis of functioning of two coupled relaxtion generators'. *Zhurnal Technicheskoi Fiziki (J. Techn. Phys.)* **11**(10). (In Russian).
10. Brunnet, L., H. Chaté, and P. Manneville: 1994, 'Long–Range Order with Local Chaos in Lattices of Diffusively Coupled ODEs'. *Physica D* **78**, 141–154.
11. Cartwright, M. L.: 1948, 'Forced oscillations in nearly sinusoidal systems'. *J. Inst. Elec. Eng.* **95**, 88.
12. Cartwright, M. L. and J. E. Littlewood: 1945, 'On nonlinear differential equations of the second order'. *J. London Math. Soc.* **20**, 180–189.
13. Chaté, H.: 1994, 'Spatiotemporal intermittency regimes of the one-dimensional complex Ginzburg-Landau equation'. *Nonlinearity* **7**, 185–204.
14. Chaté, H., A. Pikovsky, and O. Rudzick: 1999, 'Forcing Oscillatory Media: Phase Kinks vs. Synchronization'. *Physica D* **131**(1-4), 17–30.
15. Cornfeld, I. P., S. V. Fomin, and Y. G. Sinai: 1982, *Ergodic Theory*. New York: Springer.
16. Cross, M. C. and P. C. Hohenberg: 1993, 'Pattern formation outside of equilibrium'. *Rev. Mod. Phys.* **65**(3), 851.
17. Daido, H.: 1986, 'Discrete-time population dynamics of interacting self-oscillators'. *Prog. Theor. Phys.* **75**(6), 1460–1463.
18. Daido, H.: 1990, 'Intrinsic Fluctuations and a Phase Transition in a class of Large Population of Interacting Oscillators'. *J. Stat. Phys.* **60**(5/6), 753–800.
19. Denjoy, A.: 1932, 'Sur les courbes défines par les équations différentielles à la syrface du tore'. *Jour. de Matheématiques Pures at Appliquées* **11**, 333–375.
20. Dykman, G. I., P. S. Landa, and Y. I. Neymark: 1991, 'Synchronizing the Chaotic Oscillations by External Force'. *Chaos, Solitons & Fractals* **1**(4), 339–353.
21. Farmer, J. D.: 1981, 'Spectral broadening of period-doubling bifurcation sequences'. *Phys. Rev. Lett* **47**(3), 179–182.
22. Fujisaka, H. and T. Yamada: 1983, 'Stability theory of synchronized motion in coupled-oscillator systems'. *Prog. Theor. Phys.* **69**(1), 32–47.
23. Gaponov, V.: 1936, 'Two coupled generators with soft self-excitation'. *Zhurnal Technicheskoi Fiziki (J. Techn. Phys.)* **6**(5). (In Russian).

24. Glass, L.: 2001, 'Synchronization and rhythmic processes in physiology'. *Nature* **410**, 277–284.
25. Glass, L. and M. C. Mackey: 1988, *From Clocks to Chaos: The Rhythms of Life*. Princeton, NJ: Princeton Univ. Press.
26. Goryachev, A. and R. Kapral: 1996, 'Spiral waves in chaotic systems'. *Phys. Rev. Lett.* **76**(10), 1619–1622.
27. Huygens, C.: 1673, *Horologium Oscillatorium*. Parisiis, France: Apud F. Muguet. English translation: *The Pendulum Clock*, Iowa State University Press, Ames, 1986.
28. Kiss, I. and J. Hudson: 2001, 'Phase synchronization and suppression of chaos through intermittency in forcing of an electrochemical oscillator'. *Phys. Rev. E* **64**, 046215.
29. Kiss, I., Y. Zhai, and J. Hudson: 2002a, 'Collective dynamics of chaotic chemical oscillators and the law of large numbers'. *Phys. Rev. Lett.* **88**(23), 238301.
30. Kiss, I., Y. Zhai, and J. Hudson: 2002b, 'Emerging coherence in a population of chemical oscillators'. *Science* **296**, 1676–1678.
31. Kocarev, L. and U. Parlitz: 1995, 'General approach for chaotic synchronization with applications to communication'. *Phys. Rev. Lett.* **74**(25), 5028–5031.
32. Kocarev, L., A. Shang, and L. O. Chua: 1993, 'Transitions in dynamical regimes by driving: a unified method of control and synchronization of chaos'. *International Journal of Bifurcation and Chaos* **3**(2), 479–483.
33. Kuramoto, Y.: 1975, 'Self-entrainment of a Population of Coupled Nonlinear Oscillators'. In: H. Araki (ed.): *International Symposium on Mathematical Problems in Theoretical Physics*. New York, p. 420.
34. Kuramoto, Y.: 1984, *Chemical Oscillations, Waves and Turbulence*. Berlin: Springer.
35. Kurths, Editor, J.: 2000, 'A focus issue on phase synchronization in chaotic systes'. Int. J. Bifurcation and Chaos.
36. Kuznetsov, Y., P. S. Landa, A. Ol'khovoi, and S. Perminov: 1985, 'Relationship Between the Amplitude Threshold of Synchronization and the entropy in stochastic self–excited systems'. *Sov. Phys. Dokl.* **30**(3), 221–222.
37. Landa, P. S.: 1980, *Self–Oscillations in Systems with Finite Number of Degrees of Freedom*. Moscow: Nauka. (In Russian).
38. Landa, P. S. and M. G. Rosenblum: 1992, 'Synchronization of Random Self–Oscillating Systems'. *Sov. Phys. Dokl.* **37**(5), 237–239.
39. Landa, P. S. and M. G. Rosenblum: 1993, 'Synchronization and Chaotization of Oscillations in Coupled Self–Oscillating Systems'. *Applied Mechanics Reviews* **46**(7), 414–426.
40. Mandelshtam, L. and N. Papaleksi: 1947, 'On the n-th Kind Resonance Phenomena'. In: *Collected Works by L.I. Mandelshtam*, Vol. 2. Moscow: Izd. Akademii Nauk, pp. 13–20. (in Russian).
41. Mayer, A.: 1935, 'On the theory of coupled vibrations of two self-excited generators'. *Technical physics of the USSR* **11**.
42. Osipov, G., A. Pikovsky, M. Rosenblum, and J. Kurths: 1997, 'Phase Synchronization Effects in a Lattice of Nonidentical Rössler Oscillators'. *Phys. Rev. E* **55**(3), 2353–2361.
43. Ott, E.: 1992, *Chaos in Dynamical Systems*. Cambridge: Cambridge Univ. Press.
44. Panter, P.: 1965, *Modulation, Noise, and Spectral Analysis*. New York: McGraw–Hill.
45. Parlitz, U., L. Junge, W. Lauterborn, and L. Kocarev: 1996, 'Experimental Observation of Phase Synchronization'. *Phys. Rev. E.* **54**(2), 2115–2118.

46. Pecora, L. M. and T. L. Carroll: 1990, 'Synchronization in chaotic systems'. *Phys. Rev. Lett.* **64**, 821–824.

47. Pecora, Editor, L.: 1997, 'A focus issue on synchronization in chaotic systes'. CHAOS.

48. Pikovsky, A., M. Rosenblum, and J. Kurths: 1996, 'Synchronization in a Population of Globally Coupled Chaotic Oscillators'. *Europhys. Lett.* **34**(3), 165–170.

49. Pikovsky, A., M. Rosenblum, and J. Kurths: 2001, *Synchronization. A Universal Concept in Nonlinear Sciences.* Cambridge: Cambridge University Press.

50. Pikovsky, A., M. Rosenblum, G. Osipov, and J. Kurths: 1997, 'Phase Synchronization of Chaotic Oscillators by External Driving'. *Physica D* **104**, 219–238.

51. Pikovsky, A. S.: 1984, 'On the interaction of strange attractors'. *Z. Physik B* **55**(2), 149–154.

52. Pikovsky, A. S.: 1985, 'Phase synchronization of chaotic oscillations by a periodic external field'. *Sov. J. Commun. Technol. Electron.* **30**, 85.

53. Pompe, B.: 1993, 'Measuring Statistical Dependencies in a Time Series'. *J. Stat. Phys.* **73**, 587–610.

54. Quian Quiroga, R., A. Kraskov, T. Kreuz, and P. Grassberger: 2002, 'Performance of different synchronization measures in real data: A case study on electroencephalographic signals'. *Phys. Rev. E* **65**, 041903.

55. Risken, H. Z.: 1989, *The Fokker–Planck Equation.* Berlin: Springer.

56. Rosenblum, M., A. Pikovsky, and J. Kurths: 1996, 'Phase synchronization of chaotic oscillators'. *Phys. Rev. Lett.* **76**, 1804.

57. Rosenblum, M., A. Pikovsky, and J. Kurths: 1997a, 'Effect of Phase Synchronization in Driven and Coupled Chaotic Oscillators'. *IEEE Trans. CAS-I* **44**(10), 874–881.

58. Rosenblum, M., A. Pikovsky, and J. Kurths: 1997b, 'From Phase to Lag Synchronization in Coupled Chaotic Oscillators'. *Phys. Rev. Lett.* **78**, 4193–4196.

59. Rosenblum, M., A. Pikovsky, J. Kurths, G. Osipov, I. Kiss, and J. Hudson: 2002, 'Locking-based frequency measurement and synchronization of chaotic oscillators with complex dynamics'. *Phys. Rev. Lett.* **89**(26), 264102.

60. Rosenblum, M. G., G. I. Firsov, R. A. Kuuz, and B. Pompe: 1998, 'Human Postural Control: Force Plate Experiments and Modelling'. In: H. Kantz, J. Kurths, and G. Mayer-Kress (eds.): *Nonlinear Analysis of Physiological Data.* Berlin: Springer, pp. 283–306.

61. Sakaguchi, H., S. Shinomoto, and Y. Kuramoto: 1987, 'Local and global self-entrainments in oscillator lattices'. *Prog. Theor. Phys.* **77**(5), 1005–1010.

62. Schäfer, C., M. G. Rosenblum, J. Kurths, and H.-H. Abel: 1998, 'Heartbeat Synchronized with Ventilation'. *Nature* **392**(6673), 239–240.

63. Shraiman, B. I., A. Pumir, W. van Saarlos, P. Hohenberg, H. Chaté, and M. Holen: 1992, 'Spatiotemporal Chaos in the One-dimensional Ginzburg-Landau equation'. *Physica D* **57**, 241–248.

64. Stratonovich, R.: 1958, 'Oscillator synchronization in the presence of noise'. *Radiotechnika i Elektronika* **3**(4), 497.   (In Russian); English translation in: *Nonlinear Transformations of Stochastic Processes*, Edited by P.I. Kuznetsov, R.L. Stratonovich and V.I. Tikhonov, Pergamon Press, Oxford London, 1965, pp. 269-282.

65. Stratonovich, R. L.: 1963, *Topics in the Theory of Random Noise.* New York: Gordon and Breach.

66. Tang, D. Y. and N. R. Heckenberg: 1997, 'Synchronization of Mutually Coupled Chaotic Systems'. *Phys. Rev. E* **55**(6), 6618–6623.

67. Tass, P., M. G. Rosenblum, J. Weule, J. Kurths, A. S. Pikovsky, J. Volkmann,

A. Schnitzler, and H.-J. Freund: 1998, 'Detection of $n : m$ Phase Locking from Noisy Data: Application to Magnetoencephalography'. *Phys. Rev. Lett.* **81**(15), 3291–3294.

68. Teodorchik, K.: 1943, 'On the theory of synchronization of relaxational self-oscillations'. *Doklady Akademii Nauk SSSR (Sov. Phys. Dokl)* **40**(2), 63. (In Russian).

69. Ticos, C. M., E. Rosa Jr., W. B. Pardo, J. A. Walkenstein, and M. Monti: 2000, 'Experimental Real-Time Phase Synchronization of a Paced Chaotic Plasma Discharge'. *Phys. Rev. Lett.* **85**(14), 2929–2932.

70. van der Pol, B.: 1927, 'Forced oscillators in a circuit with non-linear resistance. (Reception with reactive triode)'. *Phil. Mag.* **3**, 64–80.

71. Voss, H. and J. Kurths: 1997, 'Reconstruction of Nonlinear Time Delay Models from Data by the Use of Optimal Transformations'. *Phys. Lett. A* **234**, 336–344.

CONTROL OF DYNAMICAL SYSTEMS VIA TIME-DELAYED FEEDBACK AND UNSTABLE CONTROLLER

K. PYRAGAS

Semiconductor Physics Institute and Vilnius Pedagogical University, Vilnius, Lithuania

Abstract

Time delayed-feedback control is an efficient method for stabilizing unstable periodic orbits of chaotic systems. The method is based on applying feedback proportional to the deviation of the current state of the system from its state one period in the past so that the control signal vanishes when the stabilization of the desired orbit is attained. A brief review of experimental implementations, applications for theoretical models, and most important modifications of the method is presented. Some recent results concerning the theory of the delayed feedback control as well as an idea of using unstable degrees of freedom in a feedback loop to avoid a well known topological limitation of the method are described in details.

1. Introduction

Control of dynamical systems is a classical subject in engineering science [1, 2]. The revived interest of physicists in this subject started with an idea of controlling chaos [3]. Wy chaotic systems are interesting objects for control theory and applications? The major key ingredient for the control of chaos is the observation that a chaotic set, on which the trajectory of the chaotic process lives, has embedded within it a large number of unstable periodic orbits (UPOs). In addition, because of ergodicity, the trajectory visits or accesses the neighborhood of each one of these periodic orbits. Some of these periodic orbits may correspond to a desired system's performance according to some criterion. The second ingredient is the realization that chaos, while signifying sensitive dependence on small changes to the current state and henceforth rendering unpredictable the system state in the long time, also implies that the system's behavior can be altered by using small

221

A. Pikovsky and Y. Maistrenko (eds.), Synchronization: Theory and Application, 221–256.

perturbations. Then the accessibility of the chaotic system to many different period orbits combined with its sensitivity to small perturbations allows for the control and manipulation of the chaotic process. These ideas stimulated a development of rich variety of new chaos control techniques (see Ref. [4] for review), among which the delayed feedback control (DFC) method [5] has gained widespread acceptance.

The DFC method is based on applying feedback proportional to the deviation of the current state of the system from its state one period in the past so that the control signal vanishes when the stabilization of the desired orbit is attained. Alternatively the DFC method is referred to as a method of time-delay autosynchronization, since the stabilization of the desired orbit manifets itself as a synchronization of the current state of the system with its delayed state. The DFC has the advantage of not requiring prior knowledge of anything but the period of the desired orbit. It is particularly convenient for fast dynamical systems since does not require the real-time computer processing. Experimental implementations, applications for theoretical models, and most important modifications of the DFC method are briefly listed below.

Experimental implementations.— The time-delayed feedback control has been successfully used in quite diverse experimental contexts including electronic chaos oscillators [6], mechanical pendulums [7], lasers [8], a gas discharge system [9, 10], a current-driven ion acoustic instability [11], a chaotic Taylor-Couette flow [12], chemical systems [13], high-power ferromagnetic resonance [14], helicopter rotor blades [15], and a cardiac system [16].

Applications for theoretical models.— The DFC method has been verified for a large number of theoretical models from different fields. Simmendinger and Hess [17] proposed an all-optical scheme based on the DFC for controlling delay-induced chaotic behavior of high-speed semiconductor lasers. The problem of stabilizing semiconductor laser arrays has been considered as well [18]. Rappel, Fenton, and Karma [19] used the DFC for stabilization of spiral waves in an excitable media as a model of cardiac tissue in order to prevent the spiral wave breakup. Konishi, Kokame, and Hirata [20] applied the DFC in a model of a car-following traffic. Batlle, Fossas, and Olivar [21] implemented the DFC in a model of buck converter. Bleich and Socolar [22] showed that the DFC can stabilize regular behavior in a paced, excitable oscillator described by Fitzhugh-Nagumo equations. Holyst, Zebrowska, and Urbanowicz [23] used the DFC to control chaos in economical model. Tsui and Jones investigated the problem of chaotic satellite attitude control [24] and constructed a feedforward neural network with the DFC to demonstarte a retrieval behavior that is analogous to the act of recognition [25].

The problem of controlling chaotic solitons by a time-delayed feedback mechanism has been considered by Fronczak and Holyst [26]. Mensour and Longtin [27] proposed to use the DFC in order to store information in delay-differential equations. Galvanetto [28] demonstrated the delayed feedback control of chaotic systems with dry friction. Lastly, Mitsubori and Aihara [29] proposed rather exotic application of the DFC, namely, the control of chaotic roll motion of a flooded ship in waves.

Modifications.—A reach variety of modifications of the DFC have been suggested in order to improve its performance. Adaptive versions of the DFC with automatic adjustment of delay time [30] and control gain [31] have been considered. Basso *et al.* [32] showed that for a Lur'e system (system represented as feedback connection of a linear dynamical part and a static nonlinearity) the DFC can be optimized by introducing into a feedback loop a linear filter with an appropriate transfer function. For spatially extended systems, various modifications based on spatially filtered signals have been considered [33]. The wave character of dynamics in some systems allows a simplification of the DFC algorithm by replacing the delay line with the spatially distributed detectors. Mausbach *et al.* [10] reported such a simplification for a ionization wave experiment in a conventional cold cathode glow discharge tube. Due to dispersion relations the delay in time is equivalent to the spatial displacement and the control signal can be constructed without use of the delay line. Socolar, Sukow, and Gauthier [34] improved an original DFC scheme by using an information from many previous states of the system. This extended DFC (EDFC) scheme achieves stabilization of UPOs with a greater degree of instability [35, 36]. The EDFC presumably is the most important modification of the DFC and it will be discussed at greater length in this paper.

The theory of the DFC is rather intricate since it involves nonlinear delay-differential equations. Even linear stability analysis of the delayed feedback systems is difficult. Some general analytical results have been obtained only recently [37, 38, 39, 40]. It has been shown that the DFC can stabilize only a certain class of periodic orbits characterized by a finite torsion. More precisely, the limitation is that any UPOs with an odd number of real Floquet multipliers (FMs) greater than unity (or with an odd number of real positive Floquet exponents (FEs)) can never be stabilized by the DFC. This statement was first proved by Ushio [37] for discrete time systems. Just *et al.* [38] and Nakajima [39] proved the same limitation for the continuous time DFC, and then this proof was extended for a wider class of delayed feedback schemes, including the EDFC [40]. Hence it seems hard to overcome this inherent limitation. Two efforts based on an oscillating feedback [41] and a half-period delay [42] have been taken to obviate this drawback. In both cases the mechanism of stabilization is rather unclear. Besides, the method of Ref. [42] is valid only for a special case of symmetric orbits. The limitation has been recently eliminated in a new modification of the DFC that does not utilize the symmetry of UPOs [43]. The key idea is to introduce into a feedback loop an additional unstable degree of freedom that changes the total number of unstable torsion-free modes to an even number. Then the idea of using unstable degrees of freedom in a feedback loop was drown on to construct a simple adaptive controller for stabilizing unknown steady states of dynamical systems [44].

Some recent theoretical results on the DFC method and the unstable controller are presented in more details in the rest of the paper. Section 2 is devoted to the theory of the DFC. We show that the main stability properties of the system controlled by time-delayed feedback can be simply derived from a leading Floquet exponent defining the system behavior under proportional feedback control (PFC). We consider the EDFC versus the PFC and derive the transcendental equation relating the Floquet spectra of these two control methods. At first we suppose that the FE for the PFC depends linearly on the control gain and derive the main stability properties of the EDFC. Then the case of nonlinear dependence is considered for the specific examples of the Rössler and Duffing systems. For these examples we discuss the problem of optimizing the parameters of the delayed feedback controller. In Section 3 the problem of stabilizing torsion-free periodic orbits is considered. We start with a simple discrete time model and show that an unstable degree of freedom introduced into a feedback loop can overcome the limitation of the DFC method. Then we propose a generalized modification of the DFC for torsion-free UPOs and demonstrate its efficiency for the Lorenz system. Section 4 is devoted to the problem of adaptive stabilization of unknown steady states of dynamical systems. We propose an adaptive controller described by ordinary differential equations and prove that the steady state can never be stabilized if the system and controller in sum have an odd number of real positive eigenvalues. We show that the adaptive stabilization of saddle-type steady states requires the presence of an unstable degree of freedom in a feedback loop. The paper is finished with conclusions presented in Section 5.

2.   Theory of time-delayed feedback control

If the equations governing the system dynamics are known the success of the DFC method can be predicted by a linear stability analysis of the desired orbit. Unfortunately, usual procedures for evaluating the Floquet exponents of such systems are rather intricate. Here we show that the main stability properties of the system controlled by time-delayed feedback can be simply derived from a leading Floquet exponent defining the system behavior under proportional feedback control [45]. As a result the optimal parameters of the delayed feedback controller can be evaluated without an explicit integration of delay-differential equations.

Several numerical methods for the linear stability analysis of time-delayed feedback systems have been developed. The main difficulty of this analysis is related to the fact that periodic solutions of such systems have an infinite number of FEs, though only several FEs with the largest real parts are relevant for stability properties. Most straightforward method for evaluating several largest FEs is described in Ref. [35]. It adapts the usual procedure of estimating the Lyapunov exponents of strange attractors [46]. This method requires a numerical integration of the variational system of delay-differential equations. Bleich and Socolar [36] devised an elegant method to obtain the stability domain of the system under EDFC in which the delay terms in variational equations are eliminated due to the Floquet theorem and the explicit integration of time-delay equations is avoided. Unfortunately, this method does not define the values of the FEs inside the stability domain and is unsuitable for optimization problems.

An approximate analytical method for estimating the FEs of time-delayed feedback systems has been developed in Refs. [38, 47]. Here as well as in Ref. [36] the delay terms in variational equations are eliminated and the Floquet problem is reduced to the system of ordinary differential equations. However, the FEs of the reduced system depend on a parameter that is a function of the unknown FEs itself. In Refs. [38, 47] the problem is solved on the assumption that the FE of the reduced system depends linearly on the parameter. This method gives a better insight into mechanism of the DFC and leads to reasonable qualitative results. Here we use a similar approach but do not employ the above linear approximation and show how to obtain the exact results. In this section we do not consider the problem of stabilizing torsion-free orbits and restrict ourselves to the UPOs that are originated from a flip bifurcation.

2.1. PROPORTIONAL VERSUS TIME-DELAYED FEEDBACK

Consider a dynamical system described by ordinary differential equations

$$\dot{x} = f(x, p, t), \qquad (1)$$

where the vector $x \in R^m$ defines the dynamical variables and p is a scalar parameter available for an external adjustment. We imagine that a scalar variable

$$y(t) = g(x(t)) \qquad (2)$$

that is a function of dynamic variables $x(t)$ can be measured as the system output. Let us suppose that at $p = 0$ the system has an UPO $x_0(t)$ that satisfies $\dot{x}_0 = f(x_0, 0, t)$ and $x_0(t+T) = x_0(t)$, where T is the period of the UPO. Here the value of the parameter p is fixed to zero without a loss of generality. To stabilize the UPO we consider two continuous time feedback techniques, the PFC and the DFC, both introduced in Ref. [5].

The PFC uses the periodic reference signal

$$y_0(t) = g(x_0(t)) \tag{3}$$

that corresponds to the system output if it would move along the desired UPO. For chaotic systems, this periodic signal can be reconstructed [5] from the chaotic output $y(t)$ by using the standard methods for extracting UPOs from chaotic time series data [48]. The control is achieved via adjusting the system parameter by a proportional feedback

$$p(t) = G[y_0(t) - y(t)], \tag{4}$$

where G is the control gain. If the stabilization is successful the feedback perturbation $p(t)$ vanishes. The experimental implementation of this method is difficult since it is not simply to reconstruct the UPO from experimental data.

More convenient for experimental implementation is the DFC method, which can be derived from the PFC by replacing the periodic reference signal $y_0(t)$ with the delayed output signal $y(t-T)$ [5]:

$$p(t) = K[y(t-T) - y(t)]. \tag{5}$$

Here we exchanged the notation of the feedback gain for K to differ it from that of the proportional feedback. The delayed feedback perturbation (5) also vanishes provided the desired UPO is stabilized. The DFC uses the delayed output $y(t-T)$ as the reference signal and the necessity of the UPO reconstruction is avoided. This feature determines the main advantage of the DFC over the PFC.

Hereafter, we consider a more general (extended) version of the delayed feedback control, the EDFC, in which a sum of states at integer multiples in the past is used [34]:

$$p(t) = K\left[(1-R)\sum_{n=1}^{\infty} R^{n-1}y(t-nT) - y(t)\right]. \tag{6}$$

The sum represents a geometric series with the parameter $|R| < 1$ that determines the relative importance of past differences. For $R = 0$ the EDFC transforms to the original DFC. The extended method is superior to the original in that it can stabilize UPOs of higher periods and with larger FEs. For experimental implementation, it is important that the infinite sum in Eq. (6) can be generated using only single time-delay element in the feedback loop.

The success of the above methods can be predicted by a linear stability analysis of the desired orbit. For the PFC method, the small deviations from the UPO $\delta x(t) = x(t) - x_0(t)$ are described by variational equation

$$\delta \dot{x} = [A(t) + GB(t)]\,\delta x, \tag{7}$$

where $A(t) = A(t+T)$ and $B(t) = B(t+T)$ are both T - periodic $m \times m$ matrices

$$A(t) = D_1 f(x_0(t), 0, t), \tag{8a}$$

$$B(t) = D_2 f(x_0(t), 0, t) \otimes Dg(x_0(t)). \tag{8b}$$

Here D_1 (D_2) denotes the vector (scalar) derivative with respect to the first (second) argument. The matrix $A(t)$ defines the stability properties of the UPO of the free system and $B(t)$ is the control matrix that contains all the details on the coupling of the control force.

Solutions of Eq. (7) can be decomposed into eigenfunctions according to the Floquet theory,

$$\delta x = \exp(\Lambda t) u(t), \quad u(t) = u(t+T), \tag{9}$$

where Λ is the FE. The spectrum of the FEs can be obtained with the help of the fundamental $m \times m$ matrix $\Phi(G, t)$ that is defined by equalities

$$\dot{\Phi}(G, t) = [A(t) + GB(t)]\,\Phi(G, t), \quad \Phi(G, 0) = I. \tag{10}$$

For any initial condition x_{in}, the solution of Eq. (7) can be expressed with this matrix, $x(t) = \Phi(G, t)x_{in}$. Combining this equality with Eq. (9) one obtains the system $[\Phi(G, T) - \exp(\Lambda T)I]\,x_{in} = 0$ that yields the desired eigensolutions. The characteristic equation for the FEs reads

$$\det [\Phi(G, T) - \exp(\Lambda T)I] = 0. \tag{11}$$

It defines m FEs Λ_j (or Floquet multipliers $\mu_j = \exp(\Lambda_j T)$), $j = 1 \ldots m$ that are the functions of the control gain G:

$$\Lambda_j = F_j(G), \quad j = 1, \ldots, m. \tag{12}$$

The values $F_j(0)$ are the FEs of the free system. By assumption, at least one FE of the free UPO has a positive real part. The PFC is successful if the real parts of all eigenvalues are negative, $\mathrm{Re}F_j(G) < 0$, $j = 1, \ldots, m$ in some interval of the parameter G.

Consider next the stability problem for the EDFC. The variational equation in this case reads

$$\delta \dot{\boldsymbol{x}} = A(t)\delta \boldsymbol{x}(t) + KB(t)$$
$$\times \left[(1-R)\sum_{n=1}^{\infty} R^{n-1}\delta \boldsymbol{x}(t-nT) - \delta \boldsymbol{x}(t) \right]. \tag{13}$$

The delay terms can be eliminated due to Eq. (9), $\delta \boldsymbol{x}(t-nT) = \exp(-n\Lambda T)\delta \boldsymbol{x}(t)$ As a result the problem reduces to the system of ordinary differential equations similar to Eq. (7)

$$\delta \dot{\boldsymbol{x}} = [A(t) + KH(\Lambda)B(t)]\,\delta \boldsymbol{x}, \tag{14}$$

where

$$H(\Lambda) = \frac{1 - \exp(-\Lambda T)}{1 - R\exp(-\Lambda T)} \tag{15}$$

is the transfer function of the extended delayed feedback controller. Eqs. (7) and (14) have the same structure defined by the matrices $A(t)$ and $B(t)$ and differ only by the value of the control gain. The equations become identical if we substitute $G = KH(\Lambda)$. The price one has to pay for the elimination of the delay terms is that the characteristic equation defining the FEs of the EDFC depends on the FEs itself:

$$\det \left[\Phi(KH(\Lambda), T) - \exp(\Lambda T)I \right] = 0. \tag{16}$$

Nevertheless, we can take advantage of the linear stability analysis for the PFC in order to predict the stability of the system controlled by time-delayed feedback. Suppose, that the functions $F_j(G)$ defining the FEs for the PFC are known. Then the FEs of the UPO controlled by time-delayed feedback can be obtained through solution of the transcendental equations

$$\Lambda = F_j(KH(\Lambda)), \quad j = 1 \ldots m. \tag{17}$$

Though a similar reduction of the EDFC variational equation has been considered previously (cf. Refs. [36, 38, 47]) here we emphasize the physical meaning of the functions $F_j(G)$, namely, these functions describe the dependence of the Floquet exponents on the control gain in the case of the PFC.

In the general case the analysis of the transcendental equations (17) is not a simple task due to several reasons. First, the analytical expressions of the functions $F_j(G)$ are usually unknown; they can be evaluated only numerically. Second, each FE of the free system $F_j(0)$ yields an infinite number of distinct FEs at $K \neq 0$; different eigenvalue branches that originate from different exponents of the free system may hybridize or cross so that the branches originating from initially stable FEs may become dominant in some intervals of the parameter K [47]. Third, the functions F_j in the proportional feedback technique are defined for the real-valued argument G, however, we may need a knowledge of these functions for the complex values of the argument $KH(\Lambda)$ when considering the solutions of Eqs. (17).

In spite of the above difficulties that may emerge generally there are many specific, practically important problems, for which the most important information on the EDFC performance can be simply extracted from Eqs. (15) and (17). Such problems cover low dimensional systems whose UPOs arise from a period doubling bifurcation.

In what follows we concentrate on special type of free orbits, namely, those that flip their neighborhood during one turn. More specifically, we consider UPOs whose leading Floquet multiplier is real and negative so that the corresponding FE obeys $\mathrm{Im}F_1(0) = \pi/T$. It means that the FE is placed on the boundary of the "Brillouin zone." Such FEs are likely to remain on the boundary under various perturbations and hence the condition $\mathrm{Im}F_1(G) = \pi/T$ holds in some finite interval of the control gain $G \in [G_{min}, G_{max}]$, $G_{min} < 0$, $G_{max} > 0$. Subsequently we shall see that the main properties of the EDFC can be extracted from the function $\mathrm{Re}F_1(G)$, with the argument G varying in the above interval.

Let us introduce the dimensionless function

$$\phi(G) = F_1(G)T - i\pi \tag{18}$$

that describes the dependence of the real part of the leading FE on the control gain G for the PFC and denote by

$$\lambda = \Lambda T - i\pi \tag{19}$$

the dimensionless FE of the EDFC shifted by the amount π along the complex axes. Then from Eqs. (15) and (17) we derive

$$\lambda = \phi(G), \tag{20a}$$

$$K = G\frac{1 + R\exp(-\phi(G))}{1 + \exp(-\phi(G))}. \tag{20b}$$

These equations define the parametric dependence λ versus K for the EDFC. Here G is treated as an independent real-valued parameter. We suppose that it varies in the interval $[G_{min}, G_{max}]$ so that the leading exponent $F_1(G)$ associated with the PFC remains on the boundary of the "Brillouin zone." Then the variables λ, K, and the function ϕ are all real-valued.

To demonstrate the benefit of Eqs. (20) let us derive the stability threshold of the UPO controlled by the extended time-delayed feedback. The stability of the periodic orbit is changed when λ reverses the sign. From

Eq. (20a) it follows that the function $\phi(G)$ has to vanish for some value $G = G_1$, $\phi(G_1) = 0$. The value of the control gain G_1 is nothing but the stability threshold of the UPO controlled by the proportional feedback. Then from Eq. (20b) one obtains the stability threshold

$$K_1 = G_1(1 + R)/2 \qquad (21)$$

for the extended time-delayed feedback. In Sections 2.3 and 2.4 we shall demonstrate how to derive other properties of the EDFC using the specific examples of chaotic systems, but first we consider general features of the EDFC for a simple example in which a linear approximation of the function $\phi(G)$ is assumed.

2.2. PROPERTIES OF THE EDFC: SIMPLE EXAMPLE

To demonstrate the main properties of the EDFC let us suppose that the function $\phi(G)$ defining the FE for the proportional feedback depends linearly on the control gain G (cf. Refs. [38, 47]),

$$\phi(G) = \lambda_0(1 - G/G_1). \qquad (22)$$

Here λ_0 denotes the dimensionless FE of the free system and G_1 is the stability threshold of the UPO controlled by proportional feedback. Substituting approximation (22) into Eq. (20) one derives the characteristic equation

$$k = (\lambda_0 - \lambda)\frac{1 + R\exp(-\lambda)}{1 + \exp(-\lambda)} \equiv \psi(\lambda) \qquad (23)$$

defining the FEs for the EDFC. Here $k = K\lambda_0/G_1$ is the renormalized control gain of the extended time-delayed feedback. The periodic orbit is stable if all the roots of Eq. (23) are in the left half-plane $\operatorname{Re}\lambda < 0$. The characteristic root-locus diagrams and the dependence $\operatorname{Re}\lambda$ versus k for two different values of the parameter R are shown in Fig. 1. The zeros and poles of $\psi(\lambda)$ function define the value of roots at $k = 0$ and $k \to \infty$, respectively. For $k = 0$ (an open loop system), there is a real-valued root $\lambda = \lambda_0 > 0$ that corresponds to the FE of the free UPO and an infinite number of the complex roots $\lambda = \ln R + i\pi n$, $n = \pm 1, \pm 3, \ldots$ in the left half-plane associated with the extended delayed feedback controller. For $k \to \infty$, the roots tend to the locations $\lambda = i\pi n$, $n = \pm 1, \pm 3, \ldots$ determined by the poles of $\psi(\lambda)$ function. For intermediate values of K, the roots can evolve by two different scenario depending on the value of the parameter R.

If R is small enough ($R < R^*$) the conjugate pair of the controller's roots $\lambda = \ln R \pm i\pi$ collide on the real axes [Fig. 1(a)]. After collision one of these roots moves along the real axes towards $-\infty$, and another

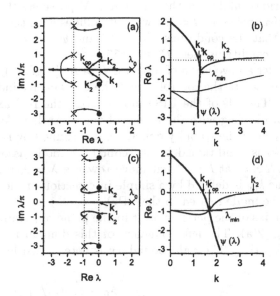

Figure 1.   Root loci of Eq. (23) as k varies from 0 to ∞ and dependence $\text{Re}\lambda$ vs. k for $\lambda_0 = 2$ and two different values of the parameter R: (a) and (b) $R = 0.2 < R^*$, (c) and (d) $R = 0.4 > R^*$. The crosses and circles denote the location of roots at $k = 0$ and $k \to \infty$, respectively. Thick solid lines in (b) and (c) symbolized by $\psi(\lambda)$ are the dependencies $k = \psi(\lambda)$ for real λ.

approaches the FE of the UPO, then collides with this FE at $k = k_{op}$ and pass to the complex plane. Afterwards this pair of complex conjugate roots move towards the points $\pm i\pi$. At $k = k_2$ they cross into the right half-plane. In the interval $k_1 < k < k_2$ all roots of Eq. (23) are in the left half-plane and the UPO controlled by the extended time-delayed feedback is stable. The left boundary of the stability domain satisfies Eq. (21). For the renormalized value of the control gain it reads

$$k_1 = \lambda_0(1 + R)/2. \tag{24}$$

An explicit analytical expression for the right boundary k_2 is unavailable. Inside the stability domain there is an optimal value of the control gain $k = k_{op}$ that for the fixed R provides the minimal value λ_{min} for the real part of the leading FE [Fig. 1(b)]. To obtain the values k_{op} and λ_{min} it suffices to examine the properties of the function $\psi(\lambda)$ for the real values of the argument λ. The values k_{op} and λ_{min} are conditioned by the maximum of this function and satisfy the equalities

$$\psi'(\lambda_{min}) = 0, \quad k_{op} = \psi(\lambda_{min}). \tag{25}$$

The above scenario is valid when the function $\psi(\lambda)$ possesses the maximum. The maximum disappears at $R = R^*$, when it collides with the minimum of this function so that the conditions $\psi'(\lambda) = 0$ and $\psi''(\lambda) = 0$ are fulfilled. For $\lambda_0 = 2$, these conditions yield $R^* \approx 0.255$.

Now we consider an evolution of roots for $R > R^*$ [Fig. 1(c),(d)]. In this case the modes related to the controller and the UPO evolve independently from each other. The FE of the UPO moves along the real axes towards $-\infty$ without hibridizating with the modes of the controller. As previously the left boundary k_1 of the stability domain is determined by Eq. (24). The right boundary k_2 is conditioned by the controller mode associated with the roots $\lambda = \ln R \pm i\pi$ at $k = 0$ that move towards $\lambda = \pm i\pi$ for $k \to \infty$. The optimal value k_{op} is defined by a simple intersection of the real part of this mode with the mode related to the UPO.

Stability domains of the periodic orbit in the plane of parameters (k, R) are shown in Fig. 2(a). The left boundary of this domain is the straight line defined by Eq. (24). The right boundary is determined by parametric equations

$$k_2 = \frac{\lambda_0^2 + s^2}{\lambda_0 + s \cot(s/2)}, \quad R = \frac{\lambda_0 - s \cot(s/2)}{\lambda_0 + s \cot(s/2)}. \tag{26}$$

with the parameter s varying in the interval $[0, \pi]$. As is seen from the figure the stability domain is smaller for the UPOs with a larger FE λ_0. Figure 2(b) shows the optimal properties of the EDFC, namely, the dependence λ_{min} versus R, where λ_{min} is the value of the leading Floquet mode evaluated at $k = k_{op}$. This dependence possesses a minimum at $R = R_{op} = R^*$. Thus for any given λ_0 there exists an optimal value of the parameter $R = R_{op}$ that at $k = k_{op}$ provides the fastest convergence of nearby trajectories to the desired periodic orbit. For $R > R_{op}$, the performance of the EDFC is adversely affected with the increase of R since for R close to 1 the modes of the controller are damped out very slowly, $\mathrm{Re}\lambda = \ln R$.

In this section we used an explicit analytical expression for the function $\phi(G)$ when analyzing the stability properties of the UPO controlled by the extended time-delayed feedback. In the next sections we consider a situation when the function $\phi(G)$ is available only numerically and only for real values of the parameter G. We show that in this case the main stability characteristics of the system controlled by time-delayed feedback can be derived as well.

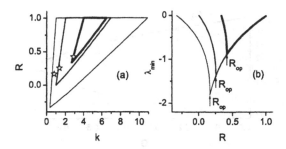

Figure 2. (a) Stability domains of Eq. (23) in (k, R) plane and (b) dependence λ_{min} vs. R for different values of λ_0: 1, 2, and 4 (increasing line thickness corresponds to increasing values of λ_0). The stars inside the stability domains denote the optimal values (k_{op}, R_{op}).

2.3. RÖSSLER SYSTEM

Let us consider the problem of stabilizing the period-one UPO of the Rössler system [49]:

$$\begin{pmatrix} \dot{x}_1 \\ \dot{x}_2 \\ \dot{x}_3 \end{pmatrix} = \begin{pmatrix} -x_2 - x_3 \\ x_1 + ax_2 \\ b + (x_1 - c)x_3 \end{pmatrix} + p(t) \begin{pmatrix} 0 \\ 1 \\ 0 \end{pmatrix}. \tag{27}$$

Here we suppose that the feedback perturbation $p(t)$ is applied only to the second equation of the Rössler system and the dynamic variable x_2 is an observable available at the system output, i.e., $y(t) = g(\boldsymbol{x}(t)) = x_2(t)$.

For parameter values $a = 0.2$, $b = 0.2$, and $c = 5.7$, the free $(p(t) \equiv 0)$ Rössler system exhibits chaotic behavior. An approximate period of the period-one UPO $\boldsymbol{x}_0(t) = \boldsymbol{x}_0(t + T)$ embedded in chaotic attractor is $T \approx 5.88$. Linearizing Eq. (27) around the UPO one obtains explicit expressions for the matrices $A(t)$ and $B(t)$ defined in Eq. (8):

$$A(t) = \begin{pmatrix} 0 & -1 & -1 \\ 1 & a & 0 \\ x_3^0(t) & 0 & x_1^0(t) - c \end{pmatrix} \tag{28}$$

and $B = \text{diag}(0, -1, 0)$. Here $x_j^0(t)$ denotes the j component of the UPO.

First we consider the system (27) controlled by proportional feedback, when the perturbation $p(t)$ is defined by Eq. (4). By solving Eqs. (10),(11) we obtain three FEs Λ_1, Λ_2 and Λ_3 as functions of the control gain G. The real parts of these functions are presented in Fig. 3(a). The values of the FEs of the free $(G = 0)$ UPO are $\Lambda_1 T = 0.876 + i\pi$, $\Lambda_2 T = 0$, $\Lambda_3 T = -31.974 + i\pi$. Thus the first and the third FEs are located on

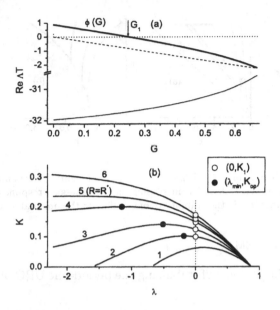

Figure 3. (a) FEs of the Rösler system under PFC as functions of the control gain G. Thick solid, thin broken, and thin solid lines represent the functions $\Lambda_1 T - i\pi$, $\Lambda_2 T$ (zero exponent), and $\Lambda_3 T - i\pi$, respectively. (b) Parametric dependence K vs. λ defined by Eqs. (20) for the EDFC. The numbers mark the curves with different values of the parameter R: (1) -0.5, (2) -0.2, (3) 0, (4) 0.2, (5) 0.28, (6) 0.4. Solid dots show the maxima of the curves and open circles indicate their intersections with the line $\lambda = 0$.

the boundary of the "Brillouin zone." The second, zero FE, is related to the translational symmetry that is general for any autonomous systems. The dependence of the FEs on the control gain G is rather complex if it would be considered in a large interval of the parameter G. In Fig. 3(a), we restricted ourselves with a small interval of the parameter $G \in [0, 0.67]$ in which all FEs do not change their imaginary parts, e.i., the FEs Λ_1 and Λ_3 remain on the boundary of the "Brillouin zone," $\mathrm{Im}\Lambda_1 T = i\pi$, $\mathrm{Im}\Lambda_3 T = i\pi$ and Λ_2 remains real-valued, $\mathrm{Im}\Lambda_2 = 0$ for any G in the above interval. An information on the behavior of the leading FE Λ_1 or, more precisely, of the real-valued function $\phi(G) = \Lambda_1 T - i\pi$ in this interval will suffice to derive the main stability properties of the system controlled by time-delayed feedback.

The main information on the EDFC performance can be gained from parametric Eqs. (20). They make possible a simple reconstruction of the relevant Floquet branch in the (K, λ) plane. This Floquet branch is shown in Fig. 3(b) for different values of the parameter R. Let us denote the dependence K versus λ corresponding to this branch by a function ψ, $K = \psi(\lambda)$. Formally, an explicit expression for this function can be written in the form

$$\psi(\lambda) = \phi^{-1}(\lambda)\frac{1 + R\exp(-\lambda)}{1 + \exp(-\lambda)}, \qquad (29)$$

where ϕ^{-1} denotes the inverse function of $\phi(G)$. More convenient for graphical representation of this dependence is, of course, the parametric form (20). The EDFC will be successful if the maximum of this function is located in the region $\lambda < 0$. Then the maximum defines the minimal value of the leading FE λ_{min} for the EDFC and $K_{op} = \psi(\lambda_{min})$ is the optimal value of the control gain at which the fastest convergence of the nearby trajectories to the desired orbit is attained. From Fig. 3(b) it is evident, that the delayed feedback controller should gain in performance through increase of the parameter R since the maximum of the $\psi(\lambda)$ function moves to the left. At $R = R^\star \approx 0.28$ the maximum disappears. For $R > R^\star$, it is difficult to predict the optimal characteristics of the EDFC. In Section 2.2 we have established that in this case the value λ_{min} is determined by the intersection of different Floquet branches.

The left boundary of the stability domain is defined by equality $K_1 = \psi(0)$ [Fig. 3(b)] or alternatively by Eq. (21), $K_1 = G_1(1 + R)/2$. This relationship between the stability thresholds of the periodic orbit controlled by the PFC and the EDFC is rather universal; it is valid for systems whose leading FE of the UPO is placed on the boundary of the "Brillouin zone." It is interesting to note that the stability threshold for the original DFC ($R = 0$) is equal to the half of the threshold in the case of the PFC, $K_1 = G_1/2$.

An evaluation of the right boundary K_2 of the stability domain is a more intricate problem. Nevertheless, for the parameter $R < R^\star$ it can be successfully solved by means of an analytical continuation of the function $\psi(\lambda)$ on the complex region. For this purpose we expand the function $\psi(\lambda)$ at the point $\lambda = \lambda_{min}$ into power series

$$\psi(\lambda) = K_{op} + \sum_{n=2}^{N+1} \alpha_n(\lambda - \lambda_{min})^n. \qquad (30)$$

The coefficients α_n we evaluate numerically by the least-squares fitting. In this procedure we use a knowledge of numerical values of the function $\psi(\lambda_m)$, $m = 1, \ldots, M$ in $M > N$ points placed on the real axes and solve a corresponding system of N linear equations. To extend the Floquet branch to the region $K > K_{op}$ we have to solve the equation $K = \psi(\lambda)$ for the complex argument λ. Substituting $\lambda - \lambda_{min} = r\exp(i\varphi)$ into Eq. (30) we obtain

$$\sum_{n=2}^{N+1} \alpha_n r^n \sin n\varphi = 0, \tag{31a}$$

$$K = K_{op} + \sum_{n=2}^{N+1} \alpha_n r^n \cos n\varphi, \tag{31b}$$

$$\mathrm{Re}\lambda = \lambda_{min} + r\cos\varphi, \tag{31c}$$

$$\mathrm{Im}\lambda = r\sin\varphi. \tag{31d}$$

Let us suppose that r is an independent parameter. By solving Eq. (31a) we can determine φ as a function of r, $\varphi = \varphi(r)$. Then Eqs. (31b),(31c) and (31b),(31d) define the parametric dependencies $\mathrm{Re}\lambda$ versus K and $\mathrm{Im}\lambda$ versus K, respectively.

Figure 4 shows the dependence of the leading FEs on the control gain K for the EDFC. The thick solid line represents the most important Floquet branch that conditions the main stability properties of the system. It is described by the function $K = \psi(\lambda)$ with the real argument λ. Note that the same function has been depicted in Fig. 3(b) for inverted axes. For $R < R^\star$, this branch originates an additional sub-branch, which starts at the point (K_{op}, λ_{min}) and spreads to the region $K > K_{op}$. The sub-branch is described by Eqs. (31) that results from an analytical continuation of the function $\psi(\lambda)$ on the complex plane. This sub-branch is leading in the region $K > K_{op}$ and its intersections with the line $\lambda = 0$ defines the right boundary K_2 of the stability domain. In Figs. 4(a),(b) the sub-branches are shown by solid lines. As seen from the figures the Floquet sub-branches obtained by means of an analytical continuation are in good agreement with the "exact" solutions evaluated from the complete system of Eqs. (10),(15),(16).

For $R > R^\star$, the maximum in the function $\psi(\lambda)$ disappears and the Floquet branch originated from the eigenvalues $\lambda = \ln R \pm i\pi$ of the controller (see Section 2.2) becomes dominant in the region $K > K_{op}$. This Floquet branch as well as the intersection point (K_{op}, λ_{min}) are unpredictable via a simple analysis. It can be determined by solving the complete system of Eqs. (10),(15),(16). In Figs. 4(c),(d) these solutions are shown by dots.

Figure 5 demonstrates how much of information one can gain via a simple analysis of parametric Eqs. (20). These equations allows us to construct the stability domain in the (K, R) plane almost completely. The most important information on optimal properties of the EDFC can be obtained from these equations as well. The thick curve in the stability domain shows the dependence of optimal value of the control gain K_{op} on the parameter R. The star marks an optimal choice of both parameters (K_{op}, R_{op}), which provide the fastest decay of perturbations. Figure 5(b) shows how the decay

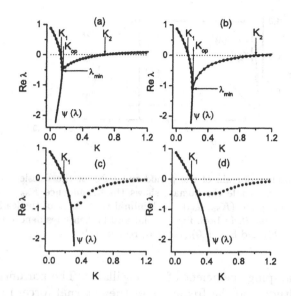

Figure 4. Leading FEs of the Rösler system under EDFC as functions of the control gain K for different values of the parameter R: (a) 0.1, (b) 0.2, (c) 0.4, (d) 0.6. Thick solid lines symbolized by $\psi(\lambda)$ show the dependence $K = \psi(\lambda)$ for real λ. Solid lines in the region $K > K_{op}$ are defined from Eqs. (31). The number of terms in series (30) is $N = 15$. Solid black dots denote the "exact" solutions obtained from complete system of Eqs. (10),(15),(16).

rate λ_{min} attained at the optimal value of the control gain K_{op} depends on the parameter R. The left part of this dependence is simply defined by the maximum of the function $\psi(\lambda)$ while the right part is determined by intersection of different Floquet branches and can be evaluated only with the complete system of Eqs. (10),(15),(16). Unlike the simple model considered in Section 2.2 here the intersection occurs before the maximum in the function $\psi(\lambda)$ disappears, i.e., at $R = R_{op} < R^\star$. Nevertheless, the value $R^\star$ gives a good estimate for the optimal value of the parameter R, since $R^\star$ is close to R_{op}.

2.4. DUFFING OSCILLATOR

To justify the universality of the proposed method we demonstrate its suitability for nonautonomous systems. As a typical example of such a system we consider the Duffing oscillator

$$\begin{pmatrix} \dot{x}_1 \\ \dot{x}_2 \end{pmatrix} = \begin{pmatrix} x_2 \\ x_1 - x_1^3 - \gamma x_2 + a \sin \omega t \end{pmatrix} + p(t) \begin{pmatrix} 0 \\ 1 \end{pmatrix}. \tag{32}$$

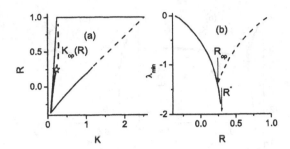

Figure 5. (a) Stability domain of the period-one UPO of the Rössler system under EDFC. The thick curve inside the domain shows the dependence K_{op} versus R. The star marks the optimal point (K_{op}, R_{op}). (b) Minimal value λ_{min} of the leading FE as a function of the parameter R. In both figures solid and broken lines denote the solutions obtained from Eqs. (20) and Eqs. (10),(15),(16), respectively.

Here γ is the damping coefficient of the oscillator. The parameters a and ω are the amplitude and the frequency of the external force, respectively. We assume that the speed x_2 of the oscillator is the observable, i.e, $y(t) = g(x(t)) = x_2$ and the feedback force $p(t)$ is applied to the second equation of the system (32). We fix the values of parameters $\gamma = 0.02$, $a = 2.5$, $\omega = 1$ so that the free $(p(t) \equiv 0)$ system is in chaotic regime. The period of the period-one UPO embedded in chaotic attractor coincides with the period of the external force $T = 2\pi/\omega = 2\pi$. Linearization of Eq. (32) around the UPO yields the matrices $A(t)$ and $B(t)$ of the form

$$A(t) = \begin{pmatrix} 0 & 1 \\ 1 - 3\left[x_1^0(t)\right]^2 & -\gamma \end{pmatrix}, \quad B = \begin{pmatrix} 0 & 0 \\ 0 & -1 \end{pmatrix}. \tag{33}$$

First we analyze the Duffing oscillator under proportional feedback defined by Eq. (4). This system is nonautonomous and does not have the zero FE. By solving Eqs. (10),(11) we obtain two FEs Λ_1 and Λ_2 as functions of the control gain G. The real parts of these functions are presented in Fig. 6(a). Both FEs of of the free $(G = 0)$ UPO are located on the boundary of the "Brillouin zone," $\Lambda_1 T = 1.248 + i\pi$, $\Lambda_2 T = 0$, $\Lambda_2 T = -1.373 + i\pi$. As before, we restrict ourselves with a small interval of the parameter $G \in [0, 1.6]$ in which both FEs remain on the boundary.

As well as in previous example the main properties of the system controlled by time-delayed feedback can be obtained from parametric Eqs. (20). Fig. 6(b) shows the dependence $K = \psi(\lambda)$ for different values of the parameter R. For the fixed value of R, the maximum of this function defines the optimal control gain $K_{op} = \psi(\lambda_{min})$. The maximum disap-

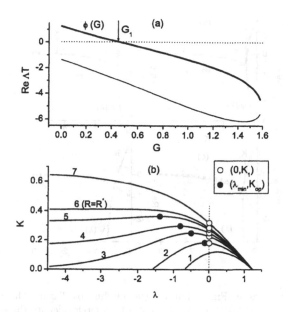

Figure 6. (a) FEs of the Duffing oscillator under PFC as functions of the control gain G. Thick and thin solid lines denote the function $\Lambda_1 T - i\pi$ and $\Lambda_2 T - i\pi$, respectively. (b) The dependence K vs. λ for the EDFC defined by parametric Eqs. (20). The numbers mark the curves with different values of the parameter R: (1) -0.5, (2) -0.2, (3) 0, (4) 0.1, (5) 0.2, (6) 0.25, (7) 0.4

pears at $R = R^\star \approx 0.25$. The left boundary of the stability domain is $K_1 = \psi(0) = G_1(1 + R)/2$, as previously.

Figure 7 shows the results of analytical continuation of the relevant Floquet branch on the region $K > K_{op}$. The continuation is performed via Eqs. (31). For small values of the parameter R [Fig. 7(a),(b)], a good quantitative agreement with the "exact" result obtained from complete system of Eqs. (10),(15),(16) is attained. For $R = 0.2 < R^\star$, the Floquet mode associated with the controller becomes dominant in the region $K > K_{op}$. In this case the analytical continuation predicts correctly the second largest FE.

Again, as in previous example, a simple analysis of parametric Eqs. 20 allows us to construct the stability domain in the (K, R) plane almost completely [Fig. 8(a)] and to obtain the most important information on the optimal properties of the delayed feedback controller [Fig. 8(b)].

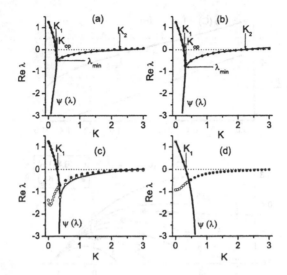

Figure 7. The same as in Fig. 4 but for the Duffing oscillator. The values of the parameter R are: (a) 0., (b) 0.1, (c) 0.2, (d) 0.4. Open circles denote the second largest FE obtained from complete system of Eqs. (10),(15),(16).

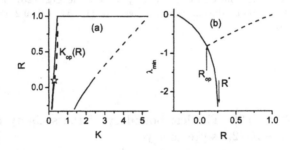

Figure 8. The same as in Fig. 5 but for the Duffing oscillator.

3. Stabilizing torsion-free periodic orbits

In Section 2 we restricted ourselves to the consideration of unstable periodic orbits erasing from a flip bifurcation. The leading Floquet multiplier of such orbits is real and negative (or corresponding FE lies on the boundary of the "Brillouin zone", $Im \Lambda = \pi/T$). Such a consideration is motivated by the fact that the usual DFC and EDFC methods work only for the orbits with a finite torsion, when the leading FE obeys $Im \Lambda \neq 0$. Unsuitability of the DFC technique to stabilize torsion-free orbits (Im$\Lambda = 0$) has been over several years considered as a main limitation of the method [37, 38, 39, 40]. More precisely, the limitation is that any UPOs with an odd number of real Floquet multipliers greater than unity can never be stabilized by the DFC. This limitation can be explained by bifurcation theory as follows. When a UPO with an odd number of real FMs greater than unity is stabilized, one of such multipliers must cross the unite circle on the real axes in the complex plane. Such a situation correspond to a tangent bifurcation, which is accompanied with a coalescence of T-periodic orbits. However, this contradicts the fact that DFC perturbation does not change the location of T-periodic orbits when the feedback gain varies, because the feedback term vanishes for T-periodic orbits.

Here we describe an unstable delayed feedback controller that can overcome the limitation. The idea is to artificially enlarge a set of real multipliers greater than unity to an even number by introducing into a feedback loop an unstable degree of freedom.

3.1. SIMPLE EXAMPLE: EDFC FOR $R > 1$

First we illustrate the idea for a simple unstable discrete time system $y_{n+1} = \mu_s y_n$, $\mu_s > 1$ controlled by the EDFC:

$$y_{n+1} = \mu_s y_n - K F_n, \tag{34}$$

$$F_n = y_n - y_{n-1} + R F_{n-1}. \tag{35}$$

The free system $y_{n+1} = \mu_s y_n$ has an unstable fixed point $y^* = 0$ with the only real eigenvalue $\mu_s > 1$ and, in accordance with the above limitation, can not be stabilized by the EDFC for any values of the feedback gain K. This is so indeed if the EDFC is stable, i.e., if the parameter R in Eq. (35) satisfies the inequality $|R| < 1$. Only this case has been considered in the literature. However, it is easy to show that the unstable controller with the parameter $R > 1$ can stabilize this system. Using the ansatz $y_n, F_n \propto \mu^n$ one obtains the characteristic equation

$$(\mu - \mu_s)(\mu - R) + K(\mu - 1) = 0 \tag{36}$$

defining the eigenvalues μ of the closed loop system (34,35). The system is stable if both roots $\mu = \mu_{1,2}$ of Eq. (36) are inside the unit circle of the μ complex plain, $|\mu_{1,2}| < 1$. Figure 1 (a) shows the characteristic root-locus diagram for $R > 1$, as the parameter K varies from 0 to ∞. For $K = 0$, there are two real eigenvalues greater than unity, $\mu_1 = \mu_s$ and $\mu_2 = R$, which correspond to two independent subsystems (34) and (35), respectively; this means that both the controlled system and controller are unstable. With the increase of K, the eigenvalues approach each other on the real axes, then

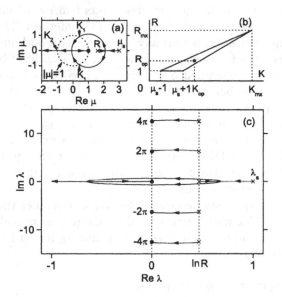

Figure 9.   Performance of (a,b) discrete and (c) continuous EDFC for $R > 1$. (a) Root loci of Eq. (36) at $\mu_s = 3$, $R = 1.6$ as K varies from 0 to ∞. (b) Stability domain of Eqs. (34,35) in the (K, R) plane; $K_{mx} = (\mu_s + 1)^2/(\mu_s - 1)$, $R_{mx} = (\mu_s + 3)/(\mu_s - 1)$. (c) Root loci of Eq. (39) at $\lambda_s = 1$, $R = 1.6$. The crosses and circles denote the location of roots at $K = 0$ and $K \to \infty$, respectively.

collide and pass to the complex plain. At $K = K_1 \equiv \mu_s R - 1$ they cross symmetrically the unite circle $|\mu| = 1$. Then both eigenvalues move inside this circle, collide again on the real axes and one of them leaves the circle at $K = K_2 \equiv (\mu_s + 1)(R + 1)/2$. In the interval $K_1 < K < K_2$, the closed loop system (34,35) is stable. By a proper choice of the parameters R and K one can stabilize the fixed point with an arbitrarily large eigenvalue μ_s. The corresponding stability domain is shown in Fig. 1 (b). For a given value μ_s, there is an optimal choice of the parameters $R = R_{op} \equiv \mu_s/(\mu_s - 1)$, $K = K_{op} \equiv \mu_s R_{op}$ leading to zero eigenvalues, $\mu_1 = \mu_2 = 0$, such that the system approaches the fixed point in finite time.

It seems attractive to apply the EDFC with the parameter $R > 1$ for continuous time systems. Unfortunately, this idea fails. As an illustration, let us consider a continuous time version of Eqs. (34,35)

$$\dot{y}(t) = \lambda_s y(t) - KF(t), \tag{37}$$

$$F(t) = y(t) - y(t - \tau) + RF(t - \tau), \tag{38}$$

where $\lambda_s > 0$ is the characteristic exponent of the free system $\dot{y} = \lambda_s y$ and τ is the delay time. By a suitable rescaling one can eliminate one of the parameters in Eqs. (37,38). Thus, without a loss of generality we can take $\tau = 1$. Equations (37,38) can be solved by the Laplace transform or simply by the substitution $y(t), F(t) \propto e^{\lambda t}$, that yields the characteristic equation:

$$1 + K \frac{1 - \exp(-\lambda)}{1 - R \exp(-\lambda)} \frac{1}{\lambda - \lambda_s} = 0. \tag{39}$$

In terms of the control theory, Eq. (39) defines the poles of the closed loop transfer function. The first and second fractions in Eq. (39) correspond to the EDFC and plant transfer functions, respectively. The closed loop system (37,38) is stable if all the roots of Eq. (39) are in the left half-plane, $\mathrm{Re}\lambda < 0$. The characteristic root-locus diagram for $R > 1$ is shown in Fig. 9 (c). When K varies from 0 to ∞, the EDFC roots move in the right half-plane from locations $\lambda = \ln R + 2\pi i n$ to $\lambda = 2\pi i n$ for $n = \pm 1, \pm 2 \ldots$. Thus, the continuous time EDFC with the parameter $R > 1$ has an infinite number of unstable degrees of freedom and many of them remain unstable in the closed loop system for any K.

3.2. USUAL EDFC SUPPLEMENTED BY AN UNSTABLE DEGREE OF FREEDOM

Hereafter, we use the usual EDFC at $0 \leq R < 1$, however introduce an additional unstable degree of freedom into a feedback loop. More specifically, for a dynamical system $\dot{x} = f(x, p)$ with a measurable scalar variable $y(t) = g(x(t))$ and an UPO of period τ at $p = 0$, we propose to adjust an available system parameter p by a feedback signal $p(t) = KF_u(t)$ of the following form:

$$F_u(t) = F(t) + w(t), \tag{40}$$

$$\dot{w}(t) = \lambda_c^0 w(t) + (\lambda_c^0 - \lambda_c^\infty)F(t), \tag{41}$$

$$F(t) = y(t) - (1 - R) \sum_{k=1}^{\infty} R^{k-1} y(t - k\tau), \tag{42}$$

where $F(t)$ is the usual EDFC described by Eq. (38) or equivalently by Eq. (42). Equation (41) defines an additional unstable degree of freedom with parameters $\lambda_c^0 > 0$ and $\lambda_c^\infty < 0$. We emphasize that whenever the stabilization is successful the variables $F(t)$ and $w(t)$ vanish, and thus vanishes the feedback force $F_u(t)$. We refer to the feedback law (40–42) as an unstable EDFC (UEDFC).

To get an insight into how the UEDFC works let us consider again the problem of stabilizing the fix point

$$\dot{y} = \lambda_s y - KF_u(t), \tag{43}$$

where $F_u(t)$ is defined by Eqs. (40–42) and $\lambda_s > 0$. Here as well as in a previous example we can take $\tau = 1$ without a loss of generality. Now the characteristic equation reads:

$$1 + KQ(\lambda) = 0, \tag{44}$$

$$Q(\lambda) \equiv \frac{\lambda - \lambda_c^\infty}{\lambda - \lambda_c^0} \frac{1 - \exp(-\lambda)}{1 - R\exp(-\lambda)} \frac{1}{\lambda - \lambda_s}. \tag{45}$$

The first fraction in Eq. (45) corresponds to the transfer function of an additional unstable degree of freedom. Root loci of Eq. (44) is shown in Fig. 10. The poles and zeros of Q-function define the value of roots at $K = 0$ and $K \to \infty$, respectively. Now at $K = 0$, the EDFC roots $\lambda = lnR + 2\pi in$, $n = 0, \pm 1, \ldots$ are in the left half-plane. The only root λ_c^0 associated with an additional unstable degree of freedom is in the right half-plane. That root and the root λ_s of the fix point collide on the real axes, pass to the complex plane and at $K = K_1$ cross into the left half-plane. For $K_1 < K < K_2$, all roots of Eq. (44) satisfy the inequality $\text{Re}\lambda < 0$, and the closed loop system (40–43) is stable. The stability is destroyed at $K = K_2$ when the EDFC roots $\lambda = \ln R \pm 2\pi i$ in the second "Brillouin zone" cross into $\text{Re}\lambda > 0$. The dependence of the five largest $\text{Re}\lambda$ on K is shown in the inset (a) of Fig. 10. The inset (b) shows the Nyquist plot, i.e., a parametric plot $\text{Re}N(\omega)$ versus $\text{Im}N(\omega)$ for $\omega \in [0, \infty]$, where $N(\omega) \equiv Q(i\omega)$. The Nyquist plot provides the simplest way of determining the stability domain; it crosses the real axes at $\text{Re}N = -1/K_1$ and $\text{Re}N = -1/K_2$.

As a more involved example let us consider the Lorenz system under the UEDFC:

$$\begin{pmatrix} \dot{x} \\ \dot{y} \\ \dot{z} \end{pmatrix} = \begin{pmatrix} -\sigma x + \sigma y \\ rx - y - xz \\ xy - bz \end{pmatrix} - KF_u(t) \begin{pmatrix} 0 \\ 1 \\ 0 \end{pmatrix}. \tag{46}$$

We assume that the output variable is y and the feedback force $F_u(t)$ [Eqs. (40–42)] perturbs only the second equation of the Lorenz system. Denote the variables of the Lorenz system by $\rho = (x, y, z)$ and those extended with the controller variable w by $\xi = (\rho, w)^T$. For the parameters $\sigma = 10$, $r = 28$, and $b = 8/3$, the free ($K = 0$) Lorenz system has a period-one UPO, $\rho_0(t) \equiv (x_0, y_0, z_0) = \rho_0(t + \tau)$, with the period $\tau \approx 1.5586$ and all real FMs: $\mu_1 \approx 4.714$, $\mu_2 = 1$ and $\mu_3 \approx 1.19 \times 10^{-10}$. This orbit can not be stabilized by usual DFC or EDFC, since only one FM is greater than unity. The ability of the UEDFC to stabilize this orbit can be verified by a linear analysis of Eqs. (46) and (40–42). Small deviations $\delta\xi = \xi - \xi_0$ from the periodic solution $\xi_0(t) \equiv (\rho_0, 0)^T = \xi_0(t + \tau)$ may be decomposed into eigenfunctions according to the Floquet theory, $\delta\xi = e^{\lambda t}u$, $u(t) = u(t + \tau)$, where λ is the Floquet exponent. The Floquet decomposition

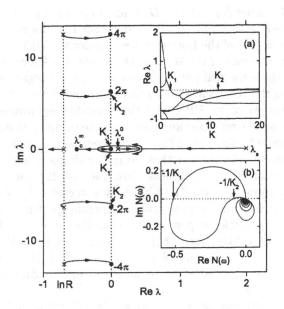

Figure 10. Root loci of Eq. (44) at $\lambda_s = 2$, $\lambda_c^0 = 0.1$, $\lambda_c^\infty = -0.5$, $R = 0.5$. The insets (a) and (b) show $\mathrm{Re}\lambda$ vs. K and the Nyquist plot, respectively. The boundaries of the stability domain are $K_1 \approx 1.95$ and $K_2 \approx 11.6$.

yields linear periodically time dependent equations $\delta\dot{\boldsymbol{\xi}} = A\delta\boldsymbol{\xi}$ with the boundary condition $\delta\boldsymbol{\xi}(\tau) = e^{\lambda\tau}\delta\boldsymbol{\xi}(0)$, where

$$A = \begin{pmatrix} -\sigma & \sigma & 0 & 0 \\ r - z_0(t) & -(1 + KH) & -x_0(t) & -K \\ y_0(t) & x_0(t) & -b & 0 \\ 0 & (\lambda_c^0 - \lambda_c^\infty)H & 0 & \lambda_c^0 \end{pmatrix}. \qquad (47)$$

Due to equality $\delta y(t - k\tau) = e^{-k\lambda\tau}\delta y(t)$, the delay terms in Eq. (42) are eliminated, and Eq. (42) is transformed to $\delta F(t) = H\delta y(t)$, where

$$H = H(\lambda) = (1 - exp(-\lambda\tau))/(1 - R\exp(-\lambda\tau)) \qquad (48)$$

is the transfer function of the EDFC. The price for this simplification is that the Jacobian A, defining the exponents λ, depends on λ itself. The eigenvalue problem may be solved with an evolution matrix Φ_t that satisfies

$$\dot{\Phi}_t = A\Phi_t, \quad \Phi_0 = I. \qquad (49)$$

The eigenvalues of Φ_τ define the desired exponents:

$$\det[\Phi_\tau(H) - e^{\lambda\tau}I] = 0. \qquad (50)$$

We emphasize the dependence Φ_τ on H conditioned by the dependence of A on H. Thus by solving Eqs. (48–50), one can define the Floquet exponents λ (or multipliers $\mu = e^{\lambda \tau}$) of the Lorenz system under the UEDFC. Figure 11 (a) shows the dependence of the six largest Reλ on K. There is an interval $K_1 < K < K_2$, where the real parts of all exponents are negative. Basically, Fig. 11 (a) shows the results similar to those presented in Fig. 10 (a). The unstable exponent λ_1 of an UPO and the unstable eigenvalue λ_c^0 of the controller collide on the real axes and pass into the complex plane providing an UPO with a finite torsion. Then this pair of complex conjugate exponents cross into domain Re$\lambda < 0$, just as they do in the simple model of Eq. (43).

Direct integration of the nonlinear Eqs. (46, 40–42) confirms the results of linear analysis. Figures 11 (b,c) show a successful stabilization of the desired UPO with an asymptotically vanishing perturbation. In this analysis, we used a restricted perturbation similar as we did in Ref. [5]. For $|F(t)| < \varepsilon$, the control force $F_u(t)$ is calculated from Eqs. (40–42), however for $|F(t)| > \varepsilon$, the control is switched off, $F_u(t) = 0$, and the unstable variable w is dropped off by replacing Eq. (41) with the relaxation equation $\dot{w} = -\lambda_r w$, $\lambda_r > 0$.

To verify the influence of fluctuations a small white noise with the spectral density $S(\omega) = a$ has been added to the r.h.s. of Eqs. (41,46). At every step of integration the variables x, y, z, and w were shifted by an amount $\sqrt{12ha}\xi_i$, where ξ_i are the random numbers uniformly distributed in the interval $[-0.5, 0.5]$ and h is the stepsize of integration. The control method works when the noise is increased up to $a \approx 0.02$. The variance of perturbation increases proportionally to the noise amplitude, $\langle F_u^2(t) \rangle = ka$, $k \approx 17$. For a large noise $a > 0.02$, the system intermittently loses the desired orbit.

4. Stabilizing and tracking unknown steady states

Although the field of controlling chaos deals mainly with the stabilization of unstable periodic orbits, the problem of stabilizing unstable steady states of dynamical systems is of great importance for various technical applications. Stabilization of a fixed point by usual methods of classical control theory requires a knowledge of its location in the phase space. However, for many complex systems (e.g., chemical or biological) the location of the fixed points, as well as exact model equations, are unknown. In this case adaptive control techniques capable of automatically locating the unknown steady state are preferable. An adaptive stabilization of a fixed point can be attained with the time-delayed feedback method [5, 35, 50]. However, the use of time-delayed signals in this problem is not necessary and thus the difficulties related to an infinite dimensional phase space due to delay can

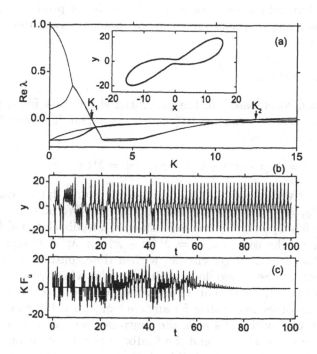

Figure 11.   Stabilizing an UPO of the Lorenz system. (a) Six largest Reλ vs. K. The boundaries of the stability domain are $K_1 \approx 2.54$ and $K_2 \approx 12.3$. The inset shows the (x,y) projection of the UPO. (b) and (c) shows the dynamics of $y(t)$ and $F_u(t)$ obtained from Eqs. (46,40–42). The parameters are: $\lambda_c^0 = 0.1$, $\lambda_c^\infty = -2$, $R = 0.7$, $K = 3.5$, $\varepsilon = 3$, $\lambda_r = 10$.

be avoided. A simpler adaptive controller for stabilizing unknown steady states can be designed on a basis of ordinary differential equations (ODEs). The simplest example of such a controller utilizes a conventional low pass filter described by one ODE. The filtered dc output signal of the system estimates the location of the fixed point, so that the difference between the actual and filtered output signals can be used as a control signal. An efficiency of such a simple controller has been demonstrated for different experimental systems [50]. Further examples involve methods which do not require knowledge of the position of the steady state but result in a nonzero control signal [51].

In this section we describe a generalized adaptive controller characterized by a system of ODEs and prove that it has a topological limitation concerning an odd number of real positive eigenvalues of the steady state [44]. We show that the limitation can be overcome by implementing an unstable degree of freedom into a feedback loop. The feedback produces a robust method of stabilizing a priori unknown unstable steady states, saddles, foci, and nodes.

4.1. SIMPLE EXAMPLE

An adaptive controller based on the conventional low-pass filter, successfully used in several experiments [50], is not universal. This can be illustrated with a simple model:

$$\dot{x} = \lambda^s(x - x^\star) + k(w - x), \quad \dot{w} = \lambda^c(w - x). \tag{51}$$

Here x is a scalar variable of an unstable one-dimensional dynamical system $\dot{x} = \lambda^s(x - x^\star)$, $\lambda^s > 0$ that we intend to stabilize. We imagine that the location of the fixed point $x^\star$ is unknown and use a feedback signal $k(w - x)$ for stabilization. The equation $\dot{w} = \lambda^c(w - x)$ for $\lambda^c < 0$ represents a conventional low-pass filter (rc circuit) with a time constant $\tau = -1/\lambda^c$. The fixed point of the closed loop system in the whole phase space of variables (x, w) is $(x^\star, x^\star)$ so that its projection on the x axes corresponds to the fixed point of the free system for any control gain k. If for some values of k the closed loop system is stable, the controller variable w converges to the steady state value $w^\star = x^\star$ and the feedback perturbation vanishes.

The closed loop system is stable if both eigenvalues of the characteristic equation $\lambda^2 - (\lambda^s + \lambda^c - k)\lambda + \lambda^s\lambda^c = 0$ are in the left half-plane $\mathrm{Re}\lambda < 0$. The stability conditions are: $k > \lambda^s + \lambda^c$, $\lambda^s\lambda^c > 0$. We see immediately that the stabilization is not possible with a conventional low-pass filter since for any $\lambda^s > 0$, $\lambda^c < 0$, we have $\lambda^s\lambda^c < 0$ and the second stability criterion is not met. However, the stabilization can be attained via an unstable controller with a positive parameter λ^c. Electronically, such a controller can be devised as the RC circuit with a negative resistance. Figure 12 shows a mechanism of stabilization. For $k = 0$, the eigenvalues are λ^s and λ^c, which correspond to the free system and free controller, respectively. With the increase of k, they approach each other on the real axes, then collide at $k = k_1$ and pass to the complex plane. At $k = k_0$ they cross symmetrically into the left half-plane (Hopf bifurcation). At $k = k_2$ we have again a collision on the real axes and then one of the roots moves towards $-\infty$ and another approaches the origin. For $k > k_0$, the closed loop system is stable. An optimal value of the control gain is k_2 since it provides the fastest convergence to the fixed point.

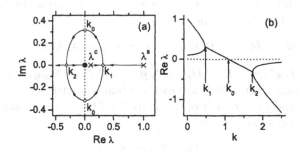

Figure 12.   Stabilizing an unstable fixed point with an unstable controller in a simple model of Eqs. (51) for $\lambda^s = 1$ and $\lambda^c = 0.1$. (a) Root loci of the characteristic equation as k varies from 0 to ∞. The crosses and solid dot denote the location of roots at $k = 0$ and $k \to \infty$, respectively. (b) Reλ vs. k. $k_0 = \lambda^s + \lambda^c$, $k_{1,2} = \lambda^s + \lambda^c \mp 2\sqrt{\lambda^s \lambda^c}$.

4.2. GENERALIZED ADAPTIVE CONTROLLER

Now we consider the problem of adaptive stabilization of a steady state in general. Let

$$\dot{x} = f(x, p) \tag{52}$$

be the dynamical system with N-dimensional vector variable x and and L-dimensional vector parameter p available for an external adjustment. Assume that an n-dimensional vector variable $y(t) = g(x(t))$ (a function of dynamical variables $x(t)$) represents the system output. Suppose that at $p = p_0$ the system has an unstable fixed point x^* that satisfies $f(x^*, p_0) = 0$. The location of the fixed point x^* is unknown. To stabilize the fixed point we perturb the parameters by an adaptive feedback

$$p(t) = p_0 + kB[Aw(t) + Cy(t)] \tag{53}$$

where w is an M-dimensional dynamical variable of the controller that satisfies

$$\dot{w}(t) = Aw + Cy. \tag{54}$$

Here A, B, and C are the matrices of dimensions $M \times M$, $M \times L$, and $n \times M$, respectively and k is a scalar parameter that defines the feedback gain. The feedback is constructed in such a way that it does not change the steady state solutions of the free system. For any k, the fixed point of the closed loop system in the whole phase space of variables $\{x, w\}$ is $\{x^*, w^*\}$, where x^* is the fixed point of the free system and w^* is the corresponding steady state value of the controller variable. The latter satisfies a system of linear equations $Aw^* = -Cg(x^*)$ that has unique solution for any nonsingular matrix A. The feedback perturbation kBw vanishes whenever the fixed point of the closed loop system is stabilized.

Small deviations $\delta x = x - x^*$ and $\delta w = w - w^*$ from the fixed point are described by variational equations

$$\delta \dot{x} = J\delta x + kPB\delta \dot{w}, \quad \delta \dot{w} = CG\delta x + A\delta w, \tag{55}$$

where $J = D_x f(x^*, p_0)$, $P = D_p f(x^*, p_0)$, and $G = D_x g(x^*)$. Here D_x and D_p denote the vector derivatives (Jacobian matrices) with respect to the variables x and parameters p, respectively. The characteristic equation for the closed loop system reads:

$$\Delta_k(\lambda) \equiv \begin{vmatrix} I\lambda - J & -k\lambda PB \\ -CG & I\lambda - A \end{vmatrix} = 0. \tag{56}$$

For $k = 0$ we have $\Delta_0(\lambda) = |I\lambda - J||I\lambda - A|$ and Eq. (56) splits into two independent equations $|I\lambda - J| = 0$ and $|I\lambda - A| = 0$ that define N eigenvalues of the free system $\lambda = \lambda_j^s$, $j = 1, \ldots, N$ and M eigenvalues of the free controller $\lambda = \lambda_j^c$, $j = 1, \ldots, M$, respectively. By assumption, at least one eigenvalue of the free system is in the right half-plane. The closed loop system is stabilized in an interval of the control gain k for which all eigenvalues of Eq. (56) are in the left half-plane $\mathrm{Re}\lambda < 0$.

The following theorem defines an important topological limitation of the above adaptive controller. It is similar to the Nakajima theorem [39] concerning the limitation of the time-delayed feedback controller.

Theorem.—Consider a fixed point x^* of a dynamical system (52) characterized by Jacobian matrix J and an adaptive controller (54) with a nonsingular matrix A. If the total number of real positive eigenvalues of the matrices J and A is odd, then the closed loop system described by Eqs. (52)-(54) cannot be stabilized by any choice of matrices A, B, C and control gain k.

Proof.—The stability of the closed loop system is determined by the roots of $\Delta_k(\lambda)$. Writing Eq. (56) for $k = 0$ in the basis where martcies J and A are diagonal, we have

$$\Delta_0(\lambda) = \prod_{j=1}^{N}(\lambda - \lambda_j^s) \prod_{m=1}^{M}(\lambda - \lambda_m^c). \tag{57}$$

Here λ_j^s and λ_m^c are the eigenvalues of the matrices J and A, respectively. Now from Eq. (56), we also have $\Delta_k(0) = \Delta_0(0)$, so Eq. (57) implies

$$\Delta_k(0) = \prod_{j=1}^{N}(-\lambda_j^s) \prod_{m=1}^{M}(-\lambda_m^c) \tag{58}$$

for all k. Since the total number of eigenvalues λ_j^s and λ_m^c that are real and positive is odd and other eigenvalues are real and negative or come in complex conjugate pairs, $\Delta_k(0)$ must be real and negative. On the other hand, from the definition of $\Delta_k(\lambda)$ we see immediately that when $\lambda \to \infty$ then $\Delta_k(\lambda) \to \lambda^{N+M} > 0$ for all k. $\Delta_k(\lambda)$ is an $N + M$ order polynomial with real coefficients and is continuous for all λ. Since $\Delta_k(\lambda)$ is negative for $\lambda = 0$ and is positive for large λ, it follows that $\Delta_k(\lambda) = 0$ for some real positive λ. Thus the closed loop system always has at least one real positive eigenvalue and cannot be stabilized, Q.E.D.

This limitation can be explained by bifurcation theory, similar to Ref. [39]. If a fixed point with an odd total number of real positive eigenvalues is stabilized, one of such eigenvalues must cross into the left half-plane on the real axes accompanied with a coalescence of fixed points. However, this contradicts the fact that the feedback perturbation does not change locations of fixed points.

From this theorem it follows that any fixed point x^* with an odd number of real positive eigenvalues cannot be stabilized with a stable controller. In other words, if the Jacobian J of a fixed point has an odd number of real positive eigenvalues then it can be stabilized only with an unstable controller whose matrix A has an odd number (at least one) of real positive eigenvalues.

4.3. CONTROLLING AN ELECTROCHEMICAL OSCILLATOR

The use of an unstable degree of freedom in a feedback loop is now demonstrated with control in an electrodissolution process, the dissolution of nickel in sulfuric acid. The main features of this process can be qualitatively described with a model proposed by Haim et al. [52]. The dimensionless model together with the controller reads:

$$\dot{e} = i - (1 - \Theta)\left[\frac{C_h \exp(0.5e)}{1 + C_h \exp(e)} + a\exp(e)\right] \tag{59a}$$

$$\Gamma\dot{\Theta} = \frac{\exp(0.5e)(1 - \Theta)}{1 + C_h \exp(e)} - \frac{bC_h \exp(2e)\Theta}{C_h c + \exp(e)} \tag{59b}$$

$$\dot{w} = \lambda^c(w - i) \tag{59c}$$

Here e is the dimensionless potential of the electrode and Θ is the surface coverage of NiO+NiOH. An observable is the current

$$i = (V_0 + \delta V - e)/R, \quad \delta V = k(i - w), \tag{60}$$

where V_0 is the circuit potential and R is the series resistance of the cell. δV is the feedback perturbation applied to the circuit potential, k is the feedback gain. From Eqs. (60) it follows that $i = (V_0 - e - kw)/(R - k)$ and $\delta V = k(V_0 - e - wR)/(R - k)$. We see that the feedback perturbation is singular at $k = R$.

In a certain interval of the circuit potential V_0, a free ($\delta V = 0$) system has three coexisting fixed points: a stable node, a saddle, and an unstable focus [Fig. 13(a)]. Depending on the initial conditions, the trajectories are attracted either to the stable node or to the stable limit cycle that surrounds an unstable focus. As is seen from Figs. 13(b) and 13(c) the coexisting saddle and the unstable focus can be stabilized with the unstable ($\lambda^c > 0$) and stable ($\lambda^c < 0$) controller, respectively if the control gain is in the interval $k_0 < k < R = 50$. Figure 13(d) shows the stability domains of these points in the (k, V_0) plane. If the value of the control gain is chosen close to $k = R$, the fixed points remain stable for all values of the potential V_0. This enables a tracking of the fixed points by fixing the control gain k and varying the potential V_0. In general a tracking algorithm requires a continuous updating of the target state and the control gain. Here described method finds the position of the steady states automatically. The method is robust enough in the examples investigated to operate without change in control gain. We also note that the stability of the saddle and focus points can be switched by a simple reversal of sign of the parameter λ^c.

Laboratory experiments for this system have been successfully carried out by I. Z. Kiss and J. L. Hudson [44]. They managed to stabilize and track both the unstable focus and the unstable saddle steady states. For the focus the usual rc circuit has been used, while the saddle point has been stabilized with the unstable controller. The robustness of the control algorithm allowed the stabilization of unstable steady states in a large parameter region. By mapping the stable and unstable phase objects the authors have visualized saddle-node and homoclinic bifurcations directly from experimental data.

5. Conclusions

The aim of this paper was to review experimental implementations, applications for theoretical models, and modifications of the time-delayed feedback control method and to present some recent theoretical ideas in this field.

In Section 2, we have demonstrated how to utilize the relationship between the Floquet spectra of the system controlled by proportional and time-delayed feedback in order to obtain the main stability properties of the system controlled by time-delayed feedback. Our consideration has been restricted to low-dimensional systems whose unstable periodic orbits are originated from a period doubling bifurcation. These orbits flip their neighborhood during one turn so that the leading Floquet exponent is placed on the boundary of the "Brillouin zone." Knowing the dependence of this exponent on the control gain for the proportional feedback control one can simply construct the relevant Floquet branch for the case of time-delayed

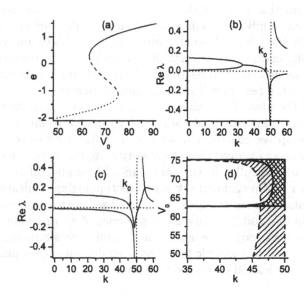

Figure 13.   Results of analysis of the electrochemical model for $R = 50$, $C_h = 1600$, $a = 0.3$, $b = 6 \times 10^{-5}$, $c = 10^{-3}$, $\Gamma = 0.01$. (a) Steady solutions e^* vs. V_0 of the free ($\delta V = 0$) system. Solid, broken, and dotted curves correspond to a stable node, a saddle, and an unstable focus, respectively. (b) and (c) Eigenvalues of the closed loop system as functions of control gain k at $V_0 = 63.888$ for the saddle $(e^*, \Theta^*) = (0, 0.0166)$ controlled by an unstable controller ($\lambda^c = 0.01$) and for the unstable focus $(e^*, \Theta^*) = (-1.7074, 0.4521)$ controlled by a stable controller ($\lambda^c = -0.01$), respectively. (d) Stability domain in (k, V_0) plane for the saddle (crossed lines) at $\lambda^c = 0.01$ and for the focus (inclined lines) at $\lambda^c = -0.01$.

feedback control. As a result the stability domain of the orbit controlled by time-delayed feedback as well as optimal properties of the delayed feedback controller can be evaluated without an explicit integration of time-delay equations. The proposed algorithm gives a better insight into how the Floquet spectrum of periodic orbits controlled by time-delayed feedback is formed. We believe that the ideas of this approach will be useful for further development of time-delayed feedback control techniques and will stimulate a search for other modifications of the method in order to gain better performance.

In Section 3 we discussed the main limitation of the delayed feedback control method, which states that the method cannot stabilize torsion-free periodic orbits, ore more precisely, orbits with an odd number of real positive Floquet exponents. We have shown that this topological limitation can be eliminated by introduction into a feedback loop an unstable degree of freedom that changes the total number of unstable torsion-free modes to an even number. An efficiency of the modified scheme has been demonstrated for the Lorenz system. Note that the stability analysis of the torsion-free orbits controlled by unstable controller can be performed in a similar manner as described in Section 2. This problem is currently under investigation and the results will be published elsewhere.

In Section 4 the idea of unstable controller has been used for the problem of stabilizing unknown steady states of dynamical systems. We have considered an adaptive controller described by a finite set of ordinary differential equations and proved that the steady state can never be stabilized if the system and controller in sum have an odd number of real positive eigenvalues. For two dimensional systems, this topological limitation states that only an unstable focus or node can be stabilized with a stable controller and stabilization of a saddle requires the presence of an unstable degree of freedom in a feedback loop. The use of the controller to stabilize and track saddle points (as well as unstable foci) has been demonstrated numerically with an electrochemical Ni dissolution system.

References

1. R. Bellman, *Introduction to the Mathematical Theory of Control Processes* (Acad. Press, New York, 1971);

2. G. Stephanopoulos, *Chemical Process Control: An Introduction to Theory and Practice* (Prentice-Hall, Englewood Cliffs, 1984);

3. E. Ott, C. Grebogi, J. A. Yorke, Phys. Rev. Lett. **64**, 1196 (1990).

4. T. Shinbrot, C. Grebogy, E. Ott, J. A. Yorke, Nature **363**, 411 (1993); T. Shinbrot, Advances in Phyzics **44**, 73 (1995); H. G. Shuster (Ed.) *Handbook of Chaos Control* (Willey-VCH, Weiheim, 1999); S. Boccaletti, C. Grebogi, Y.-C. Lai, H. Mancini, D. Maza, Physics Reports **329**, 103 (2000).

5. K. Pyragas, Phys. Lett. A **170**, 421 (1992).

6. K. Pyragas, A. Tamaševičius, Phys. Lett. A **180** 99 (1993); A. Kittel, J. Parisi, K. Pyragas, R. Richter, Z. Naturforsch. **49a** 843 (1994); D. J. Gauthier, D. W. Sukow, H. M. Concannon, J. E. S. Socolar, Phys. Rev. E **50**, 2343 (1994); P. Celka, Int. J. Bifurcation Chaos Appl. Sci. Eng. **4**, 1703 (1994).

7. T. Hikihara, T. Kawagoshi, Phys. Lett. A **211**, 29 (1996); D. J. Christini, V. In, M. L. Spano, W. L. Ditto, J. J. Collins, Phys. Rev. E **56** R3749 (1997).

8. S. Bielawski, D. Derozier, P. Glorieux, Phys. Rev. E **49**, R971 (1994); M. Basso, R. Genesio R, A. Tesi, Systems and Control Letters **31**, 287 (1997); W. Lu, D. Yu, R. G. Harrison, Int. J. Bifurcation Chaos Appl. Sci. Eng. **8**, 1769 (1998).

9. T. Pierre, G. Bonhomme, A. Atipo, Phyz. Rev. Lett. **76** 2290 (1996); E. Gravier, X. Caron, G. Bonhomme, T. Pierre, J. L. Briancon, Europ. J. Phys. D **8**, 451 (2000).

10. Th. Mausbach, Th. Klinger, A. Piel, A. Atipo, Th. Pierre, G. Bonhomme, Phys. Lett. A **228**, 373 (1997).

11. T. Fukuyama, H. Shirahama, Y. Kawai, Physics of Plasmas **9**, 4525 (2002).

12. O. Lüthje, S. Wolff, G. Pfister, Phys. Rev. Lett. **86**, 1745 (2001).

13. P. Parmananda, R. Madrigal, M. Rivera, L. Nyikos, I. Z. Kiss, V. Gaspar, Phys. Rev. E **59**, 5266 (1999); A. Guderian, A. F. Munster, M. Kraus, F. W. Schneider, J. of Phys. Chem. A **102**, 5059 (1998);

14. H. Benner, W. Just, J. Korean Pysical Society **40**, 1046 (2002).

15. J. M. Krodkiewski, J. S. Faragher, J. Sound and Vibration **234** (2000).

16. K. Hall, D. J. Christini, M. Tremblay, J. J. Collins, L. Glass, J. Billette, Phys. Rev. Lett. **78**, 4518 (1997).

17. C. Simmendinger, O. Hess, Phys. Lett. A **216**, 97 (1996).

18. M. Munkel, F. Kaiser, O. Hess, Phys. Rev. E **56**, 3868 (1997); C. Simmendinger, M. Munkel, O. Hess, Chaos, Solitons and Fractals **10**, 851 (1999).

19. W. J. Rappel, F. Fenton, A. Karma, Phys. Rev. Lett. **83**, 456 (1999).

20. K. Konishi, H. Kokame, K. Hirata, Phyz. Rev. E **60**, 4000 (1999); K. Konishi, H. Kokame, K. Hirata, European Phyzical J. B **15**, 715 (2000).

21. C. Batlle, E. Fossas, G. Olivar, Int. J. Circuit Theory and Applications **27**, 617 (1999).

22. M. E. Bleich, J. E. S. Socolar, Int. J. Bifurcation Chaos Appl. Sci. Eng. **10**, 603 (2000).

23. J. A. Holyst, K. Urbanowicz, Phsica A **287**, 587 (2000); J. A. Holyst, M. Zebrowska, K. Urbanowicz, European Phyzical J. B **20**, 531 (2001).

24. A. P. M. Tsui, A. J. Jones, Physica D **135**, 135, 41 (2000).

25. A. P. M. Tsui, A. J. Jones, Int. J. Bifurcation Chaos Appl. Sci. Eng. **9**, 713 (1999).

26. P. Fronczak, J. A. Holyst, Phys. Rev. E **65**, 026219 (2002).

27. B. Mensour, A. Longtin, Phys. Lett. A **205**, 18 (1995).

28. U. Galvanetto, Int. J. Bifurcation Chaos Appl. Sci. Eng. **12**, 1877 (2002).

29. K. Mitsubori, K. U. Aihara, Proceedings of the Royal Society of London Series A-Mathematical Phyzics and Engeneering Science **458**, 2801 (2002).

30. K. Pyragas, Phys. Lett. A **198**, 433 (1995); H. Nakajima, H. Ito, Y. Ueda, IE-ICE Transactions on Fundamentals of Electronics Communications and Computer Sciencies, **E80A**, 1554, (1997); G. Herrmann, Phys. Lett. A **287**, 245 (2001).

31. S. Boccaletti, F. T. Arecchi, Europhys. Lett. **31**, 127 (1995); S. Boccaletti, A. Farini, F. T. Arecchi, Chaos, Solitons and Fractals **8**, 1431 (1997).

32. M. Basso, R. Genesio, A. Tesi, IEEE Trans. Circuits Syst. I **44**, 1023 (1997); M. Basso, R. Genesio, L Giovanardi, A. Tesi, G. Torrini, Int. J. Bifurcation Chaos Appl. Sci. Eng. **8**, 1699 (1998).

33. M. E. Bleich, D. Hochheiser, J. V. Moloney, J. E. S. Socolar, Phys. Rev. E **55**, 2119 (1997); D. Hochheiser, J. V. Moloney, J. Lega, Phys. Rev. A **55**, R4011 (1997); N. Baba, A. Amann, E. Scholl, W. Just, Phys. Rev. Lett. **89**, 074101 (2002).

34. J. E. S. Socolar, D. W. Sukow, D. J. Gauthier, Phys. Rev. E **50**, 3245 (1994).

35. K. Pyragas, Phys. Lett. A **206**, 323 (1995).

36. M. E. Bleich, J. E. S. Socolar, Phys. Lett. A **210**, 87 (1996).

37. T. Ushio, IEEE Trans. Circuits Syst. I **43**, 815 (1996).

38. W. Just, T. Bernard, M. Ostheimer, E. Reibold, H. Benner, Phys. Rev. Lett. **78**, 203 (1997).

39. H. Nakajima, Phys. Lett. A **232**, 207 (1997).

40. H. Nakajima, Y. Ueda, Physica D **111**, 143 (1998).

41. S. Bielawsky, D. Derozier, P. Glorieux, Phys. Rev. A **47**, R2492 (1993); H.G. Shuster, M.B. Stemmler, Phys. Rev. E **56**, 6410 (1997).

42. H. Nakajima, Y. Ueda, Phys. Rev. E **58**, 1757 (1998).

43. K. Pyragas, Phys. Rev. Lett. **86**, 2265 (2001).

44. K. Pyragas, V. Pyragas, I. Z. Kiss, J. L. Hudson, Phys. Rev. Lett. **89**, 244103 (2002).

45. K. Pyragas, Phys. Rev. E **66**, 026207 (2002).
46. G. Benettin, C. Froeschle, J. P. Scheidecker, Phys. Rev. A **19**, 2454 (1979); I. Schimada, T. Nagashima, Prog. Theor. Phys. **61**, 1605 (1979).
47. W. Just, E. Reibold, H. Benner, K. Kacperski, P. Fronczak, J. Holyst, Phys. Lett. A **254**, 158 (1999); W. Just, E. Reibold, K. Kacperski, P. Fronczak, J. A. Holyst, H. Benner, Phys. Rev. E **61**, 5045 (2000).
48. D. P. Lathrop, E. J. Kostelich, Phys. Rev. A **40**, 4028 (1989); P. So, E. Ott, S. J. Schiff, D. T. Kaplan, T. Sauer, C. Grebogi, Phys. Rev. Lett. **76**, 4705 (1996); P. So, E. Ott, T. Sauer, B. J. Gluckman, C. Grebogi, S. J. Schiff, Phys. Rev. E **55**, 5398 (1997).
49. O. E. Rössler, Phys. Lett. A **57**, 397 (1976).
50. A. Namajūnas, K. Pyragas, A. Tamaševičius, Phys. Lett. A **204**, 255 (1995); N. F. Rulkov, L. S. Tsimring, H. D. I. Abarbanel, Phys. Rev. E **50**, 314 (1994); A. S. Z. Schweinsberg, U. Dressler, Phys. Rev. E **63**, 056210 (2001).
51. E. C. Zimmermann, M. Schell, J. Ross, J. Chem. Phys. **81**, 1327 (1984); J. Kramer, J. Ross, J. Chem. Phys. **83**, 6234 (1985); B. Macke, J. Zemmouri, N. E. Fettouhi, Phys. Rev. A **47**, R1609 (1993).
52. D. Haim, O. Lev, L. M. Pismen, M. J. Sheintuch, J. Phys. Chem., **96**, 2676 (1992).

Index